# Palgrave Studies in Media and Environmental Communication

Series Editors
Anders Hansen
Department of Media and Communication
University of Leicester
Leicester, UK

Steve Depoe
McMicken College of Arts and Sciences
University of Cincinnati
Cincinnati, OH, USA

D1333787

Drawing on both leading and emerging scholars of environmental communication, the Palgrave Studies in Media and Environmental Communication Series features books on the key roles of media and communication processes in relation to a broad range of global as well as national/local environmental issues, crises and disasters. Characteristic of the cross-disciplinary nature of environmental communication, the books showcase a broad variety of theories, methods and perspectives for the study of media and communication processes regarding the environment. Common to these is the endeavour to describe, analyse, understand and explain the centrality of media and communication processes to public and political action on the environment.

**Advisory Board**
Stuart Allan, Cardiff University, UK
Alison Anderson, Plymouth University, UK
Anabela Carvalho, Universidade do Minho, Portugal
Robert Cox, The University of North Carolina at Chapel Hill, USA
Geoffrey Craig, University of Kent, UK
Julie Doyle, University of Brighton, UK
Shiv Ganesh, Massey University, New Zealand
Libby Lester, University of Tasmania, Australia
Laura Lindenfeld, University of Maine, USA
Pieter Maeseele, University of Antwerp, Belgium
Chris Russill, Carleton University, Canada
Joe Smith, The Open University, UK


More information about this series at
http://www.springer.com/series/14612

Benedetta Brevini · Graham Murdock
Editors

# Carbon Capitalism and Communication

Confronting Climate Crisis

*Editors*
Benedetta Brevini
School of Literature, Arts and Media
University of Sydney
Sydney, NSW
Australia

Graham Murdock
Department of Social Sciences
Loughborough University
Loughborough, Leicestershire
UK

Palgrave Studies in Media and Environmental Communication
ISBN 978-3-319-57875-0       ISBN 978-3-319-57876-7   (eBook)
DOI 10.1007/978-3-319-57876-7

Library of Congress Control Number: 2017944592

Cover illustration: Design Pics/Don Hammond

Printed on acid-free paper

This Palgrave Macmillan imprint is published by Springer Nature
The registered company is Springer International Publishing AG
The registered company address is: Gewerbestrasse 11, 6330 Cham, Switzerland

# CONTENTS

# EDITORS AND CONTRIBUTORS

## About the Editors

**Dr. Benedetta Brevini** is a journalist, media activist and Senior Lecturer at the University of Sydney. Before joining academia she worked as journalist in Milan, New York and London for CNBC and RAI. She writes on *The Guardian's* Comment is Free and contributes to a number of print and web publications including the *Conversation, Opendemocracy, Index of Censorship* and *Red Pepper Magazine*. She is the author of *Public Service Broadcasting* online (2013) and editor of the acclaimed volume *Beyond Wikileaks* (2013).

**Prof. Graham Murdock** is Professor of Culture and Economy at Loughborough University and Vice President of the International Association of Media and Communication Research. He has been translated into twenty one languages and has held Visiting Professor and Fellow positions in Australia, Belgium, China, New Zealand, Norway and Sweden. His recent publications include; *Money Talks: Media, Markets, Crisis* with Jostein Gripsrud (2015); *The Handbook of Political Economy of Communications* with Janet Wasko and Helena Sousa (2011); and *Digital Dynamics: Engagement and Disconnections* with Peter Golding (2010).

## Contributors

**Dr. Jo Bates** research examines the socio-material factors that influence the production and use of data, and that enable and restrict the movement of data between different people and organisations. This includes examining how government policies and legislation shape how data move between different sites of practice.

More recently, Dr. Bates led an AHRC project (The Secret Life of a Weather Datum) which examined the socio-cultural values, practices and public policies shaping the journey and form of meteorological data from its initial production through to being re-used in different contexts, including climate science and financial markets.

Source: https://www.sheffield.ac.uk/is/staff/batesresearch

**Professor Jodi Dean** is a political philosopher and Professor in the Political Science department at Hobart and William Smith Colleges. She has also held the position of Erasmus Professor of the Humanities in the Faculty of Philosophy at Erasmus University Rotterdam.

Drawing from Marxism, psychoanalysis, post-structuralism, and post-modernism, Professor Dean has made contributions to contemporary political theory, media theory, and feminist theory, most notably with her theory of communicative capitalism; the online merging of democracy and capitalism into a single neoliberal formation that subverts the democratic impulses of the masses by valuing emotional expression over logical discourse.

She is the co-editor of the journal *Theory & Event*.

Source: https://en.wikipedia.org/wiki/Jodi_Dean

**Dr. Shane Gunster** is an Associate Professor and the Graduate Program Chair in the School of Communication at Simon Fraser University. His research and teaching interests focus upon environmental communication, especially the politics of climate and energy. His recent work has been published in the *Canadian Journal of Communication*, the *Canadian Journal of Political Science* and edited collections with Wilfrid Laurier University Press, University of Toronto Press and MIT Press. He is currently working on a book manuscript on environmental journalism. He has also worked closely with the Canadian Centre for Policy Alternatives on a variety of research projects associated with the Climate Justice Project.

Source: http://pics.uvic.ca/events/good-life-green-life

**Professor Robert Hackett** is Professor of Communication, Simon Fraser University. He has written extensively on media democratization, and journalism as political communication. His most recent collaborative books include *Expanding Peace Journalism: Comparative and Critical Approaches* (2011), and *Remaking Media: The Struggle to Democratize Public Communication* (2006). He is on the editorial advisory board of *Journalism Studies, Journal of Alternative and Community Media*, and other academic journals. He has co-founded several community-oriented media education and advocacy initiatives, including NewsWatch Canada, OpenMedia.ca, and Media Democracy Days.
Source: https://www.sfu.ca/communication/people/faculty/hackett.html

**Dr. Mitchell Hobbs** is Lecturer in Media and Public Relations at the University of Sydney. His research activities concern political communication and media power, and his publications are regularly assigned as core texts at universities in Australia and the United States. Mitchell also possesses high-level experience in media and public relations. Most notably, he worked in political public relations for Prime Minister Julia Gillard from 2011 to 2012. In this role, he was responsible for the implementation of the Hon Julia Gillard's media and communication activities in her electorate in Melbourne. Mitchell's professional experiences and research activities have given him unique insights into communication power and its capacity for social and political change.
Source: https://sydney.edu.au/arts/media_communications/staff/profiles/mitchell.hobbs.php

**Naomi Klein** is an award-winning journalist, syndicated columnist and author of the international bestsellers, *This Changes Everything: Capitalism versus The Climate* (2014), *The Shock Doctrine: The Rise of Disaster Capitalism* (2007) and *No Logo* (2000).

In 2017, Klein became Senior Correspondent for *The Intercept*. She is also a Puffin Foundation Writing Fellow at The Nation Institute and contributor to the *Nation Magazine*. Recent articles have also appeared in *The New York Times, The New Yorker, The Boston Globe, The Guardian*, the *London Review of Books* and *Le Monde*.

She has multiple honourary degrees and in 2014 received the International Studies Association's IPE Outstanding Activist-Scholar award.
Source: http://www.naomiklein.org/meet-naomi

**Professor Justin Lewis** is Professor of Communication at Cardiff School of Journalism, Media and Cultural Studies, and Dean of Research for the College of Arts, Humanities and Social Sciences.

He has written widely about media, culture and politics. His books, since 2000, include *Constructing Public Opinion* (2001), *Citizens or Consumers: What the media tell us about political participation* (2005), *Shoot First and Ask Questions Later: Media Coverage of the War in Iraq* (2006), *Climate Change and the Media* (2009) and *The world of 24 hour news* (2010). *His latest book is Beyond Consumer Capitalism: Media and the Limits to Imagination* (2013). He has also written books on media audiences, cultural policy and media and race.
Source: http://www.cardiff.ac.uk/people/view/182947-lewis-justin

**Dr. Michael E. Mann**  is Distinguished Professor of Atmospheric Science at Penn State University, with joint appointments in the Department of Geosciences and the Earth and Environmental Systems Institute (EESI). He is also director of the Penn State Earth System Science Center (ESSC). His research involves the use of theoretical models and observational data to better understand Earth's climate system.

Dr. Mann is author of more than 200 peer-reviewed and edited publications, and has published three books including *Dire Predictions: Understanding Climate Change* (2008), *The Hockey Stick and the Climate Wars: Dispatches from the Front Lines* (2012), and most recently, *The Madhouse Effect* (2016) with *Washington Post* editorial cartoonist Tom Toles.    Source: http://www.michaelmann.net/content/about
Source: http://www.michaelmann.net/content/about

**Professor Richard Maxwell** is a political economist of media. His research begins at the intersection of politics and economics to analyze the global media, their social and cultural impact, and the policies that regulate their reach and operations. He has published widely on a range of topics, from television in Spain's democratic transition to Hollywood's international dominance, from media politics in the post 9–11 era to how big political economic forces work in the mundane routines of daily life and culture.

His writing on media and cultural consumption draws attention to the specter of living life under ever expanding governmental and commercial surveillance. His current work on the environmental impact of media focuses on the environmental harms caused by media, information technologies, and electronics.
Source: http://www.qc.cuny.edu/Academics/Degrees/DAH/Media Studies/Pages/RichardMaxwell.aspx

**Dr. Patrick McCurdy** is an Associate Professor in the Department Communication at the University of Ottawa and a Core Member of the University of Ottawa's Institute for Science, Society and Policy. His research draws from media and communication, journalism, social movement studies and the energy humanities to examine media as a site and source of social struggle and contestation. Most recently, his work has studied the evolution of advertising and campaigning around Alberta's oil/tar sand in his project Mediatoil (www.mediatoil.ca). Patrick's work has been published in several academic journals and he is the co-author of *Protest Camps* and the co-editor of three books: *Protest Camps in International Context: Spaces, Infrastructures and Media of Resistance* (2017), *Beyond WikiLeaks: Implications for the Future of Communications, Journalism and Society* (2013) and *Mediation and Protest Movements* (2013).
Source: https://arts.uottawa.ca/communication/en/people/mccurdy-patrick

**Associate Professor David McKnight** is a researcher interested broadly in politics, history, journalism and the media. Most recently he was a co-author of *Big Coal: Australia's Dirtiest Habit* (2013). He is also the author of several books including *Rupert Murdoch: An Investigation of Political Power* (2012) and *Beyond Right and Left: New Politics and the Culture War* (2005) which discusses the renewal of the progressive political vision. He co-edited (with Robert Manne) *Goodbye To All That?*

(2010), a collection of essays on the failure of neo-liberalism. In 2012 he was the co-author of *Journalism at the Speed of Bytes*, a study commissioned by the Walkley Foundation for Journalism on the future of journalism in view of the crisis in the business model of newspapers.
Source: https://www.arts.unsw.edu.au/about-us/people/david-mcknight/

**Professor Toby Miller** is the Professor and Director of the Institute for Media and Creative Industries at Loughborough University London.

He is vastly experienced in this field and has numerous publications in Television; New Media, Sport; Culture, Cultural Politics and Film Theory. Prior to pursuing academics, Professor Miller worked in broadcasting, banking, and the civil service.

Professor Miller has recently published a book entitled *Global Media Studies* with Marwan M. Kraidy, the Director of the Center for Advanced Research in Global Communication at the Annenberg School for Communication at the University of Pennsylvania.
Source: http://www.lborolondon.ac.uk/about/staff/toby-miller/

**Emeritus Professor Vincent Mosco** is Professor Emeritus of Sociology at Queen's University and Distinguished Professor of Communication, New Media Centre, Fudan University, Shanghai. At Queen's he was Canada Research Chair in Communication and Society and head of the Department of Sociology. His research interests include the political economy of communication, the social impacts of information technology, and communication policy. Dr. Mosco is the author or editor of 21 books including *The Digital Sublime* (2004) and *The Political Economy of Communication* (2009). His *To the Cloud: Big Data in a Turbulent World*, was named a 2014 Outstanding Academic Title by Choice: Current Reviews for Academic Libraries. He is currently working on *Becoming Digital: Toward a Post-Internet Society*.
Source: http://www.vincentmosco.com/

**Blair Palese's** experience is in people and project management, strategy development and communications across a range of eco and social issues, particularly climate change and marine protection. She is the CEO of 350. org in Australia working to build a movement to stop new fossil fuels and rapidly reduce greenhouse emissions. She has worked in Australia, the

United Kingdom and the United States in senior management roles for Greenpeace International, The Body Shop internationally, and as a consultant to a range of companies, organisations and government projects.

She is currently on the board of Greenpeace Australia Pacific and was a founding committee member for Human Rights Watch Australia.
Source: https://www.linkedin.com/in/blair-palese-9b43435/

**David Ritter** is the CEO of Greenpeace Australia Pacific.

David returned to Australia to take up this role in 2012 after five years working in a senior campaigns position with Greenpeace in London. There he worked on the global issues of destructive fishing, deforestation and climate change. Prior to joining Greenpeace, David worked as an academic and a lawyer in both commercial and native title practices.

David is a widely published commentator on politics, law, history and current affairs.

He is an Honorary Fellow of the Faculty of Law of the University of Western Australia and an Associate of the Institute for Democracy and Human Rights at the University of Sydney.
Source: https://www.linkedin.com/in/david-ritter-5b813717/

**Alan Rusbridger** was editor-in-chief of *Guardian News & Media* from 1995 to 2015. He is member of The Scott Trust, which owns *The Guardian* and *The Observer*.

As editor, he helped launch *Guardian Unlimited*—now theguardian.com—and, in 2004, was responsible for the paper's complete redesign and transformation into the European Berliner format. He oversaw the integration of the paper and digital operations, helping to build a website which today attracts visits from more than 100 million unique browsers a month. Now the world's second largest serious newspaper website, it has regularly been voted the best newspaper website in the world.

Rusbridger has been named editor of the year three times. He has won the Liberty Human Rights Award, the European Press Prize and the Ortega y Gasset award and has been honoured by CUNY, Columbia, Oslo and Syracuse Universities.
Source: https://www.theguardian.com/global/2010/aug/26/alan-rusbridger-profile

**Professor Kim Sheehan** brings more than 12 years of experience in advertising and marketing to the School of Journalism and Communication's (SOJC) advertising sequence. She has held leadership positions at agencies in Boston, Chicago, and St. Louis and has consulted for numerous companies, including People magazine and Laura Ashley. She is a past president of the American Academy of Advertising. She has written or edited 11 books and has just signed a contract to write *The Academic's Guide to Mechanical Turk*.

Sheehan, who is a past director of the Strategic Communication Master's program at the SOJC's George S. Turnbull Portland Center, teaches courses in advertising and brand planning, media theory, and research methodology. She also directs the SOJC's Honors Program and works with many graduate students.

Source: http://journalism.uoregon.edu/member/sheehan_kim/

**Professor Xin Tong** is an Associate Professor of Urban and Economic Geography at Peking University. With a background in economic geography, or industrial geography more specifically, she is currently interested in the spatial dimension of industrial ecology, both theoretical and practical.

Professor Xin Tong has worked on the Extended Producer Responsibility in e-waste management in China, especially the response from the domestic industry at this policy strategy at national, regional and local levels for years, now extending her research interests to cover the relationship between global environmental governance and technological change in developing countries broadly.

She also participates in consultancy projects, such as regional industrial strategic planning and industrial park planning for local governments.

Source: http://pku.academia.edu/XTong/CurriculumVitae

**Professor Christopher Wright** is Professor of Organisational Studies at the University of Sydney Business School where he teaches and researches organisational change, management innovation, sustainability and critical understandings of capitalism and political economy.

His current research explores organizational and societal responses to climate change, with a particular focus on how managers and business organizations interpret and respond to the climate crisis. He has published

on this topic in relation to issues of corporate environmentalism, corporate citizenship, organizational justification and compromise, risk, identity and future imaginings. His research on climate change and business is internationally recognised and he has developed research collaborations with leading international climate scientists and global environmental organisations.
Source: https://sydney.edu.au/business/staff/christopher.wright

**Dr. Terry Woronov** Since completing an undergraduate degree in Chinese Studies at Georgetown University, Terry Woronov has studied social and political change during China's long twentieth century. She has lived, worked, and studied in China, Hong Kong and Taiwan for many years, including spending 18 months in Beijing (1999–2001) for her Ph.D. fieldwork research, and a year in Nanjing (2007–2008) for a subsequent fieldwork project in a vocational secondary school.

Dr. Woronov holds a Ph.D. in Anthropology from the University of Chicago, an M.Ed. from Harvard, and was a post-doctoral research fellow at the Institute for East Asian Studies at the University of California-Berkeley.
Source:    https://sydney.edu.au/arts/anthropology/staff/profiles/terry.woronov.php

# LIST OF FIGURES

# CHAPTER 1

# Carbon, Capitalism, Communication

*Graham Murdock and Benedetta Brevini*

## CARBON

Despite being written almost 200 years ago, Mary Shelley's 1818 novel *Frankenstein* remains the best known cautionary tale of the risks and damage that may follow from human intervention in fundamental natural processes. It opens with a series of letters home from a ship's captain whose vessel is trapped in the Arctic. Looking out from the deck he sees "stretched out in every direction, vast and irregular plains of ice, which seemed to have no end" (Shelley 1992: 25 [1818]). The next day the crew rescues the forlorn figure of Frankenstein, who is pursuing the man-monster he has created in order to kill him. Almost 200 years later, the story has retained a central place in the popular imagination. But Mary Shelley's vision of an endless sheet of white stretching to the horizon has been replaced by iconic photographs of polar bears clinging to small slivers of ice, surrounded by dark ocean, anchoring the impact of climate change in a powerfully resonant image.

G. Murdock
Loughborough University, Loughborough, UK

B. Brevini (✉)
University of Sydney, Sydney, Australia
e-mail: benedetta.brevini@sydney.edu.au

© The Author(s) 2017
B. Brevini and G. Murdock (eds.), *Carbon Capitalism and Communication*,
Palgrave Studies in Media and Environmental Communication,
DOI 10.1007/978-3-319-57876-7_1

The Arctic ice mass plays a key role in regulating global temperatures, reflecting the sun's radiation back into space and cooling the winds that blow over it and the water that passes under it. In November 2016, temperatures in the Arctic were around 20 °C warmer than expected for that time of year and sea ice coverage was the lowest ever recorded. The release of these findings coincided with the publication of the *Arctic Resilience Report*, the most comprehensive scientific study of the region to date, lending support to the authors' conclusion that continuing rapid melting of the polar ice cap, and consequent warming, could trigger tipping points and feedback loops that will have major impacts on global climate patterns, not only in the far north but in the heavily populated middle latitudes (Arctic Council 2016). One key tipping point is reached when the warming of offshore permafrost releases increased volumes of methane, a major greenhouse gas, causing extreme weather, disrupted food production and increased health risks. One calculation, simply for the thawing of permafrost beneath the East Siberian Sea, puts the total cost of these impacts at \$60 trillion, only \$10 trillion short of the estimated size of the world economy in 2012. As the authors note, "The total cost of Arctic change will be much higher" (Whiteman et al. 2013: 401). At the same time, the prospect of ice-free passage through polar waters and the opportunity to access the 30% of the world's undiscovered gas and 13% of undiscovered oil in the Arctic are powerful incentives for further corporate exploitation.

Recent severe weather conditions provide a stark reminder that the disruption caused by accelerating interruptions to natural cycles is not confined to remote regions or future dates. They are here now. Figures for global average near surface temperatures (measured a metre above ground level) confirm that 2016 overtook 2015 as the warmest year recorded since 1850, with 90% of the increase attributable to the high levels of carbon dioxide in the atmosphere, levels not for seen for 4 million years (Met Office 2017). The consequences are far-reaching with seal levels rising and "climate-related extremes such as heat waves, heavy precipitation and droughts increasing in frequency and intensity" disrupting food production, aiding the spread of diseases previously confined to the tropics, and accelerating species extinctions (European Environmental Agency 2017: 12).

The latest research confirms consequences that were already evident a decade ago and had prompted the Nobel Prize winning atmospheric chemist Paul Crutzen and his colleagues to argue that: "Human activities have become so pervasive and profound that they are pushing the Earth into planetary *terra incognita*. The Earth is rapidly moving into a less biologically diverse, less forested, much warmer, and probably wetter and stormier state" (Steffen et al. 2007: 614).

They see these escalating impacts pointing to a step-change in the relation between human activity and the earth's system and suggest that we have now entered a new "interval of time on Earth in which many key processes are dominated by human influence" (Zalasiewicz et al. 2011: 835). In an earlier paper 'Geology of Mankind' published in the influential scientific journal *Nature*, Crutzen named this new period the Anthropocene. Drawing on observations of the increasing concentrations of the two major 'greenhouse' gases contributing to global warming, carbon dioxide and methane, in the air trapped in samples of polar ice, he located the start of the new era "in the latter part of eighteenth century" a date he notes, "happened to coincide with James Watt's design of a steam engine in 1784" (Crutzen 2002: 23). The contribution of Watt's invention to the development of early industrialisation is currently the focus of dispute among historians (Malm 2016) but by placing it centre stage in his account of the dynamics propelling the rise of the Anthropocene, Crutzen raises two key issues.

Firstly, while the mechanisation of industrial production and transportation clearly did accelerate the shift in energy generation from renewable sources to finite and exhaustible reserves of fossil fuels, initially coal and later oil and natural gas, we must take care not to conflate 'renewable' with 'clean'. While water and wind power entail minimal emissions, all forms of carbon-based energy contribute to the production of the major greenhouse gas, carbon dioxide, with major impacts on global warming and ocean acidification. Recent action on climate change has tended to focus on reducing reliance on fossil fuels. A notable example is *The Guardian's* 'Keep it in the Ground' campaign described by Alan Rusbridger in this collection. In looking for alternatives, there is mounting support for policies that address future energy needs by increasing the use of biosolids and liquids derived from trees or recently harvested plants. There are two problems with this approach. Firstly, when burned in the presence of oxygen, wood, particularly wood pellets, currently the dominant form of solid biomass traded internationally, usually emit "more greenhouse gases per unit of energy produced than fossil fuels", to which we need to add the "supply-chain emissions from harvesting, collecting, processing and transport". As a consequence "the use of woody biomass for energy will release higher levels of emissions than coal and considerably higher levels than gas"

(Brack 2017: 9–10). Secondly, while biofuels such as ethanol derived from fermenting maize, sugarcane or wheat offer superficially attractive alternatives to gasoline, as a recent report from the World Resources Institute points out, "the dedicated use of land to generate bioenergy...is likely to compete with food production and carbon storage", making feeding the planet more difficult and increasing greenhouse emissions (Searchinger and Heimlich 2015: 26).

Writers on the Anthropocene have taken the volume of carbon dioxide ($CO_2$) in the atmosphere as the signature index of climate change. In 1750 the atmospheric concentration of $CO_2$ stood at 277 parts per million (ppm) by volume. By 1850 it had climbed to 285 ppm and by 1945 it was 25 ppm above the pre-industrial level, well outside the range of natural variability displayed during the long era since the last ice age, the Holocene, and providing "the first indisputable evidence that human activities were affecting the environment on a global scale" (Steffen et al. 2007: 616). The inclusion of 'carbon' rather than 'fossil fuels' in the title of this collection reflects the centrality of this argument.

Crutzen's decision to date the start of the Anthropocene from the onset of the industrial revolution in Britain also underlines the need to place issues around climate change firmly in the context of fundamental transformations in the organisation of production and consumption, and the shifts in resource extraction and energy generation they required. In a later paper, he adds deforestation and cattle farming to fossil fuel burning as major contributors to the escalation of $CO_2$ and other major greenhouse gases in the atmosphere. In the same paper he proposes a modification to his original time line, arguing that that there has been a very significant intensification in the impact of human intervention since 1950 and that this second phase of the Anthropocene "stands out as the one in which human activities rapidly changed from merely *influencing* the global environment in some ways to *dominating* it in many ways", to the extent that "Human impacts on the Earth System structure and functioning now equal or exceed in magnitude many forces of nature at the global scale" (Crutzen and Steffen 2003: 253). Compelling evidence for this 'Great Acceleration' has come from the comprehensive analysis of temperature changes over the last six centuries conducted by Michael E. Mann (who contributes to this present collection) and his colleagues, and published in *Nature* in 1998 (Mann et al. 1998). Using proxy measures of change, including tree rings and coral growth, to generate data for periods before formal observations began, the research revealed a hockey stick pattern with a relatively stable long historic line of modest variations suddenly ending in a sharp, almost vertical, rise, the blade, with three of the 8 years prior to publication being warmer than for any time since 1400. A later paper, taking a longer time frame, confirmed their central argument that while past changes in climate could be plausibly attributed to shifts in solar and volcanic activity "neither can explain the dramatic warming of the late twentieth century" and "only anthropogenic influences (principally the increases in greenhouse gas concentrations)" provide a causal explanation (Jones and Mann 2004: 31).

The argument that global warming was primarily driven by emissions from cars, factories, power plants and other core components of life in advanced industrial societies attracted a barrage of hostility from climate change deniers wedded to business as usual. Mann was subjected to sustained attacks on his professional credibility and personal threats. The Republican attorney general of Virginia pressed to have his academic credentials revoked and in 2009 emails he had exchanged with British climate scientist Phil Jones were stolen and posted in selective and distorted form on the internet, weeks before the United Nations climate talks were due to begin in Copenhagen. Mann sees their release as a part of a deliberate effort to undermine the talks, which ended in failure, an outcome he describes as a "crime against humanity, a crime against the planet" (McKie 2012). As he points out in the interview included in this collection, although successive investigations into the emails completely exonerated him, they took several years, during which time climate change deniers were able to repeatedly exploit the doubts raised about his findings. Other attacks were more personal, with emails sent to his in box demanding that "You and your colleagues... be shot, quartered and fed to the pigs along with your whole damn families," or "hopin [sic] I would see the news and you commited [sic] suicide" (McKie 2012).

While Mann's research confirms the 'great acceleration' in climate change since 1950, it also points to a more recent, second acceleration. In a recent paper Paul Crutzen and his colleagues have identified the years between 1950 and 1973 as a distinctive period (Steffen et al. 2011: 850), and argued that human influence on the climate has been most evident in the years since 1970 (Gaffney and Steffen 2017: 4). To make sense of these dates we need to place debates around the Anthropocene firmly in the context of shifts in the organisation of capitalism. As Jason Moore has argued:

"The Anthropocene makes for an easy story to tell..... The mosaic of human activity in the web of life is reduced to an abstract Humanity, a homogenous acting unit. Inequality, commodification, imperialism…and much more, have been largely removed from consideration" (Moore 2015: 170).

Writing on the Anthropocene tends to focus on the extension of industrialisation rather than transformation of capitalism. While industrialisation outside the orbit of capitalism, most obviously in the Soviet Union, has played a role in the acceleration of climate change, the major contributors to greenhouse gases since 1750 have been the capitalist economies of Western Europe and North America. With the collapse of the Soviet Union, the turn to the market in China and India and the rise of South Korea and other emerging economies, variants of capitalism have for the first time achieved global reach. This has led commentators to argue that we need to talk about the 'Capitalocene' rather than the 'Anthropocene' (see Moore 2016) and examine how the changing organisation of capitalism has shaped the systems of extraction, production and consumption that are propelling climate change.

Moving capitalism to the centre of the analysis also forces us to confront the unequal responsibilities and impacts of climate change. "The dominant narrative of the Anthropocene presents an abstract humanity uniformly involved-and, it implies, uniformity to blame" (Bonneuil and Fressoz 2016: 66). Rising rates of personal consumption have contributed to climate crisis, but the major damage has been done by corporations relentlessly pursuing profit generation at the minimum cost. The consequences have born down hardest on those with the least resources to protect or insulate themselves. The accelerating destruction of habitats and native lands, and the exhaustion of scarce natural resources has been central to a process of accumulation that has progressively concentrated wealth in the hands of a diminishing group of super-rich captains of capitalism. The heads of digital communication companies now figure prominently in this elite group. The current *Forbes* list of the world's leading billionaires has Bill Gates of Microsoft in first place, Jeff Bezos of Amazon in third place, Mark Zuckerberg of Facebook in fifth place, with Larry Ellison of Oracle and Larry Page and Sergey Brin of Google in the top thirteen (Forbes 2017). The movement of communication companies to the centre of contemporary capitalism and the battle over climate crisis is the end result of a long process.

## CAPITALISM

The periodisation emerging from scientific research on the Anthropocene coincides with three distinct phases in the development of contemporary capitalism. At each stage the impacts on climate change and the natural environmental have intensified as new developments in the organisation of production and consumption have added to the cumulative store of greenhouse gases and the increasing encroachment of extraction and agriculture on forests and wilderness.

The initial phase, covering the century between 1850 and 1950 saw the rapid extension and consolidation of industrialised production across the leading capitalist economies of Europe and North America together with rising levels of personal consumption and increasing state provision of essential services. These developments relied overwhelmingly on energy generated from fossil fuels, coal and later oil and natural gas, significantly increasing the amount of carbon dioxide in the atmosphere, which, as we noted earlier, proponents of the Anthropocene have taken as its signature index. The rise of industrial capitalism also fundamentally altered the population balance between rural areas and urban concentrations, prompting an accelerated mechanisation of agricultural production to meet the challenge of ensuring adequate food supplies for the rapidly expanding industrial conurbations. Between 1800 and 1950 the amount of the earth's surface that was 'domesticated' increased from around 10% to 25–30% (Steffen et al. 2007: 616). Wilderness and natural habitats contracted, and deforestation accelerated the significant expansion of meat production, which increased the volume of the other major greenhouse gas, methane, released into the atmosphere.

The second phase, between 1950 and 1973, saw three major developments. Firstly, the escalating consumption of standardised industrial products, dubbed Fordism after Henry Ford, whose model T automobile had become the iconic image of rising consumer aspirations, established itself as the dominant model of capitalist organisation in the countries of Western Europe, which had previously lagged behind the United States. At the end of World War II there were 40 million motor vehicles globally. By 1996 that figure had risen to 700 million. Emissions were further boosted by the rapid increase in international air travel. Images of the pleasures and comforts of increased consumption, militantly promoted by an expanded advertising industry, were increasing disseminated on a global basis, raising expectations and reorienting conceptions of the good life around consumer choice. Initially however, the economies emerging in the newly independent former colonies of the major western powers were oriented to rapid industrialisation. The transformation of countries like South Korea into major sites of industrial production, with the associated consequences for rising emissions, constituted the second major shift in the organisation of capitalism in the post-war period. The third was the advent of nuclear power and nuclear weapons, with the associated rise in levels of radioactive particles from the series of tests conducted by the major powers.

In the mid-1970s, the advanced capitalist societies experienced a major crisis of profitability. This opened a space that was increasing commandeered by market fundamentalist rhetoric that identified the causes of crisis with the regime of regulation and state management installed in the post war period. This had imposed curbs on corporate exploitation and boosted the provision of public resources as key components of a social settlement aimed at guaranteeing rights of citizenship rooted in full and equal social participation. The counter argument that resource allocation was best decided by corporations responding to market disciplines found powerful political advocates in Margaret Thatcher and Ronald Reagan, and their elections as heads of government in two of the leading capitalist economies provided policy templates for a fundamental restructuring of both the corporate system and the organisation of production and consumption that was adopted in other major centres of capitalism.

Public assets were sold to private investors and previously monopoly or protected markets were opened to competition, substantially expanding the scope of corporate activity. At the same time, many of the regulations that had imposed public interest restraints on these activities were removed or diluted, giving companies significantly more flexibility in their operations. The easing of restrictions on mergers and acquisitions led to concerted waves of consolidation in a number of sectors, reinforcing the political leverage that the major corporations could exert. Substantial volumes of industrial production were moved overseas to take advantage of the lower wage rates paid to labour in emerging economies. Many of these offshore plants and factories relied on energy from low grade coal with high levels of pollution. This was particularly evident in China where market-oriented reforms had been introduced following the fall of Mao Tse-tung, and where atmospheric pollution has emerged as a central focus of public protest.

Concerted efforts to lower production costs were accompanied by a major reengineering of social identities, moving away from the idea of citizens as members of moral and social communities with responsibilities to contribute to the quality of collective life and installing the identity of consumers, pursuing personal satisfaction and advancement through market choices, as the dominant social imaginary. Consumers were encouraged to purchase more products more often, internalize the logic of rapid obsolescence and disposability and reject the ethos of retention and repair that had characterised relations to big ticket items in the post-war period. These shifts were propelled by a massive extension of personal debt as credit and store cards displaced cash payments, and by a concerted push to present consumer goods as preeminent arenas of self-expression and self-realisation. The utilitarian appeals to convenience and value for money that had underpinned sales of mass consumer durables in the period between 1950 and 1973 were progressively overtaken by a promotional culture that emphasised the unique qualities of particular brands and cemented their associations with desirable life styles. Increasingly, as incomes in emerging economies rose, this new consumer system achieved global currency. At the end of 2016, eight of the world's ten largest shopping malls as measured by 'gross leasable area' were in Asia (Touropia 2016).

These shifts in the organisation of production and consumption generated rising demands for both energy and essential materials, prompting a marked escalation in extractive activities and a concerted push to identify and exploit hitherto untapped reserves, moving into areas that had previously been untouched and utilizing new, and potentially hazardous, methods of extraction, such as fracking.

Much contemporary debate has been underpinned by a "narrative of ecological awakening", which sees awareness of impending climate crisis as relatively recent (Bonneuil and Fressoz 2016: 171). This ignores the long history of concern dating back to the early phase of industrial capitalism.

In 1857, a provincial French lawyer, Eugene Huzar, published *The Tree of Science*, drawing attention to the finite nature of the planet's resources. "For us, our planet is limited, very limited. ...When one sees something as limited as the earth, and a power as unlimited as man using science, one can only wonder what impact this power will have, one day, on our poor small earth" (quoted in Fessoz 2007: 335). He warned against the dire environmental consequences of capitalism's coal fuelled pursuit of profit, writing "As man becomes more involved with industry and uses more coal you can predict that in one or two centuries, the world being criss-crossed by railroads and steamboats and being covered by factories and industrial plants, billions of tons of carbon dioxide will be emitted, and as the forests will have been eliminated, these billions of tons of carbon could well trouble a little the world's harmony" (quoted in Fessoz 2007: 335–336).

Despite his amateur status, Huzar was widely read (Fessoz 2007). But at the time there was no firm scientific backing for his speculations. By the end of the century there was, with Svante Arrhenius and Thomas Chamberlin publishing papers within a year of each other, in 1896 and 1897, offering detailed explorations of the relationship between accumulations of $CO_2$ in the atmosphere and global warming (Zalasiewicz et al. 2011: 833). The captains of capitalism were well aware of this but chose to carry on regardless, denying the evidence, exploiting disagreements within the scientific community, and transferring the social and environmental costs of climate change from their balance sheets to the public purse. As a number of the contributions to this volume demonstrate in detail, these strategies are still very much in play. Even when fossil fuel companies have acknowledged that global warming is accelerating they have chosen to pursue business as usual. In 1991, the major oil corporation, Shell, released *Climate of Concern*, a short film designed to be shown in schools and universities, as part of the company's continuing effort to massage their public image and establish their green credentials. The film warned that although "global warming is not yet certain, many think that to wait for final proof would be irresponsible" and went on to argue that "action now is seen as the only safe insurance" against increased incidents of extreme weather, floods, famines and climate refugees (quoted in Carrington and Mommers 2017: 3). Despite this declared position, and its own research showing that exploiting new reserves of oil and gas comprehensively undermined the goal of controlling global warming, the company continued investing in tar sands, advocating fracking and pressing for permission to drill in the Arctic.

## COMMUNICATION

Shell's release of *Climate of Concern* as part of a wider public relations strategy points to the importance of the media system as the key arena in which competing narratives and arguments struggle for visibility and legitimacy. The chapters included in this collection offer a range of case studies detailing the strategies and tactics employed on both sides of this battle, from the corporate advocacy backing the continued expansion of coal mining in Australia to the mobilisation of celebrity support and endorsement in the struggle over the exploitation of the tar sands in Canada. As well as drawing on recent research, we have asked a number of prominent activists involved in combating climate crisis to reflect on their communication strategies.

The pivotal role of communication systems in sifting and presenting information, orchestrating public debate, crafting resonant images and engaging stories that frame public anxieties and concerns, apportion responsibility, and advocate action is a key link in the chain connecting the operation of communications under capitalism to the climate crisis. These resources for understanding, and misunderstanding, are displayed in full public view and the work of journalists and other media professionals, and the corporations and institutions that employ them, are often newsworthy in their own right. But, there is a second, equally pivotal, link between the organisation of communication and climate crisis, which is less visible but no less important that has received far less attention. One of the main aims of this present volume is to bring these two connecting links together.

The production and consumption of public information and imagery depends on a dense array of communication infrastructures and machines, from underground cables and satellites to widescreen television sets, laptops, tablets and smart phones. Up until comparatively recently, the implications of the very obvious fact that media systems and equipment are assembled from a range of natural and synthetic materials, consume energy, produce emissions in their production and use and contribute to problems of pollution and waste in their disposal has attracted surprisingly little comment or analysis. "In communication and media scholarship, the overwhelming focus has been on texts, the industry that produces them, and the viewers that consume them; the materiality of devices and networks has been consistently overlooked" (Gillespie et al. 2014: 1). The material presence of media constitutes a backstory that has remained mostly untold, a hinterland that has remained largely unmapped until very recently. Two contributors to this volume, Rick Maxwell and Tony Miller, have been in the forefront of efforts to provide a map, and their contribution here summarises the present debate. Recognising the material presence of communication systems alongside their central role as key spaces of advertising and pivotal arenas for constructing and contesting public understandings is essential to a complete account of the intersections between developments in media, transformations in capitalism and the escalation of the climate crisis.

The expansion of the media system in the initial phase of capitalism's consolidation, between 1850 and 1950, can usefully be divided into two phases. The first phase, from 1850 to the turn of the twentieth century, was dominated by two developments: the rapid growth of the popular commercial press and affordable printed novels, and the introduction and expansion of the telegraph network.

The telegraph was the first technology to separate communications from transportation. Previously, messages had to be conveyed in physical form, as a letter, card or gift. With the universal adoption of Samuel Morse's coding system of dots and dashes, they could transmitted as digital pulses over a wired network assuming material form only when they were decoded and written out by the operator who received them. At the same time this system continued to rely on the physical networks of roads, rail links and ocean routes that transported the raw materials required to construct and operate telegraph links and deliver messages to their addressees.

This basic principle holds true for contemporary digital media where, as with the telegraph, communication is released from its physical casings and translated into digital files. This has persuaded some commentators to characterise digital media as weightless and immaterial, pointing to the displacement of printed books and newspapers by e-books and websites, and the substitution of streamed access to recorded music, films and television programmes for physical disc storage. There are two very obvious problems with his assumption. Firstly, the production of digital media still involves the deployment of machines and spaces that consume material resources and energy. The self-publishing author of an e-book working from home is using a computer and printer, probably storing drafts in the cloud, consuming power and relying on a physical network to reach readers. Secondly, those readers can only access the text if they have a laptop, smart phone or dedicated e-book reader, again very tangible artefacts that are assembled from complex combinations of materials and production sites, rely on physical networks, and require access to energy to power them

The history of the telegraph also reminds us that communication systems have come to play an increasingly central role in the co-ordination of geographically dispersed corporate and governmental activities. Despite its formal openness to anyone, cost rapidly tipped regular use of the telegraph towards institutional rather than individual users. Consequently, addressing the role of communication systems in exacerbating the climate crisis requires concerted action on corporate and state uses of media networks and machines alongside action on personal consumption.

The second phase in the consolidation of capitalism, from 1900 to 1950, saw a new complex of communication facilities become firmly embedded in everyday life. On the one hand, cinema established itself as the preeminent form of popular leisure, facilitating the collective consumption of resources and energy. On the other hand, a new array of media goods, gramophones, telephones, radios and Kodak cameras promoted domestic and personalised consumption, multiplying both the volume of communication goods in circulation and the demands they made on resources and energy.

These innovations in media played a double role in the consolidation of capitalism. They significantly increased the range of commodities and services that could be sold to consumers and, in the case of radio in the US and other countries that rejected non-commercial, public service forms of organisation, provided an extended platform for advertising that promoted other consumer goods and reinforced a mode of address that hailed listeners as consumers rather than citizens.

The second major phase in the development of capitalism, from 1945 to 1973 saw television replace cinema as the primary focus of popular mediated leisure and increasing mobility of access to both radio, with the invention of transistors, and recorded music, with the expansion of portable record players and later cassette tape machines. As a consequence, alongside collective cinema-going and shared television viewing around a single domestic installation, individualised consumption proliferated, expanding the demand for both materials and energy and creating increased problems of waste.

The years since 1973 have seen both mobility and individualisation rapidly increase with the arrival of the internet as a mass popular utility and the migration of access from desk-top computers to tablets and smart phones. At the same time, the space available to advertising has increased very significantly. The arrival of commercial cable, satellite and terrestrial television channels broke the monopolies previously commanded by public broadcasting organizations in China, India and across Europe, while the business models developed by the major internet companies depend on harvesting users' personal data for analysis and resale to advertisers wanting to target their appeals more accurately, and providing platforms for the development of new forms of promotion utilizing the web's interactivity to link appeals to consumerism even more securely to self-images.

In addition to reinforcing the wastefulness of over-consumption in general, embedding digital media at the centre of everyday life has had three other major impacts with consequences for climate crisis.

Firstly, the always there, always on, nature of tablet and smart phone access to the internet has significantly increased the demand on power supplies. Use is no longer periodic. It is continuous.

Secondly, smart phones have been in the forefront of efforts to develop 'frictionless' purchasing, by-passing cash and cards, and encouraging consumers to swipe their devices across payment points. The intention is to increase the volume and rate of consumption by reducing to the absolute minimum the time for second thoughts (see Justin Lewis's chapter in this volume).

Thirdly, digital media, particularly smart phones, have played a major role in accelerating rates of product obsolescence and disposal by encouraging users to upgrade on a regular basis. The latest version of the iPhone is the seventh model to be launched, consigning the previous six to the scrapheap. The result is a mounting accumulation of electronic waste, much of it non-biodegradable and some of it toxic. As Xin Tong demonstrates in her chapter, in a variant of offshore activity, China has been a major recipient of discarded machines, where their salvage and re-use supports a thriving microeconomy with complex and uneasy relations to urban expansion and city planning. This case forcibly reminds us that behind the glossy ads promoting the latest laptops and smart phones lies an extended chain of production, use and disposal that links us as end-users to the miners extracting the minerals required for manufac-ture, the assembly workers labouring under exploitative conditions, seamen flying flags of convenience in the container ships that transport the finished machines, the staff in the coal and oil-fired power stations that produce the energy we consume in use, and the rural migrants newly arrived in the cities searching the piles of our discarded waste for anything of value. Retrieving these lives and exploring their role in the reproduction of contemporary capitalism is essential to any full analysis of the present climate crisis and ways it might be addressed.

Additionally, the digital machines we buy, use and discard come in packages that create pollution. Plastics are a particular problem. A recent review of available research by the World Economic Forum estimated that there are currently over 150 million tonnes of plastic in the world's oceans and that if business as usual continues by 2050 these will contain more plastics than fish (by weight) and that production of plastics will account for 20% of total oil consumption and 15% of the carbon budget needed to keep global warming below the internationally agreed level of 2 °C (World Economic Forum 2016: 7). Plastics are produced and disposed of in a number of other areas of economic activity, but the contribution made by communications sectors needs to be included in any full analysis of the connection between media and climate crisis.

As Vincent Mosco points out in his contribution to this volume, with the expansion of cloud computing and the rapid development of artificial intelligence, communication systems are likely to make greater calls on energy and scarce resources in future, increasing their contribution to the climate crisis. The transfer of user data from flash drives and other portable storage devices to the massive server farms that constitute the 'cloud' significantly increases demand for power to operate the facilities and water to cool them. The increasing application of robotics and artificial intelligence to productive processes and the Internet of Things, which connects an increasing range of 'smart' domestic machines and devices to communication networks that collect and analyse data they transmit on patterns of use independent of the volition and control of their owners, will again impose substantially increased demands on network capacity and power supplies.

## The Present Collection

### Section One: Contested Futures

The collection opens with interviews with two of the key contributors to contemporary debates around the causes of climate crisis and the actions required to address it.

In the opening chapter, Michael E. Mann, architect of the iconic 'hockey stick' graph, mentioned earlier, demonstrating the acceleration of global warming, reflects on the responsibility of scientists to enter the public debate and on his experience of being a target of attack from climate change deniers. His most recent work, *The Madhouse Effect*, sees him teaming up with political cartoonist Tom Toles to satirise the twisted logic of denialists.

In Chap. 2, the prominent Canadian writer and activist, Naomi Klein, whose best-selling book *This Changes Everything* (2014) laid the blame for the present climate crisis firmly at the door of unrestrained capitalism, and which has played a major role in propelling this argument to the center of contemporary debate, recounts how she came to this conclusion following the watershed events of Hurricane Katrina, when American elites capitalised on the disaster both during and after the immediate crisis. The privatisation of the transport and energy sectors have inhibited the ability of governments to control two of the economies most crucial to securing change and highlighted the endemic conflict between the need for drastic emission reductions and capitalism's imperative of constant growth. She argues that the dominant ideologies of neoliberalism and market fundamentalism have proved themselves fundamentally incompatible with the recognition of interdependence that collective action against climate change demands. In framing strategies for opposition and alternatives, she highlights the emergence of 'Blockadia', the grassroots climate activism spearheaded by indigenous and farming communities motivated by dependence on, and reverence for, the land.

## Section Two: Toxic Technologies: Media Machines and Ecological Crisis

The first three chapters in the second section explore the ways that contemporary media systems make a double contribution to the climate crisis, through the materials and energy they consume in production and use and the waste they create in disposal, and in their central role as sites of product promotion fuelling unsustainable levels of consumption.

In Chap. 3, Richard Maxwell and Toby Miller, who have been in the forefront of work underlining the need to study media as material structures and artefacts, review the available research on the environmental impact of media and communication technologies, and the challenges citizens and environmental groups face in greening their uses and ensuring pleasurable media consumption and competent citizenship in the light of electronic waste.

In Chap. 4, Justin Lewis argues that although the energy and transport industries are commonly considered as the major contributors to climate crisis, the media sector is almost uniquely destructive because its strategies for generating profit depend on creating as much electronic waste as possible through planned obsolescence. This business model relies on selling hardware with a built-in expiry date that becomes outdated and requires replacement within 2 years. Some companies have gone further by taking measures to prevent consumers from extending the life of their devices. The environmental consequences of this hyper-production cycle are crippling, with the majority of $CO_2$ emitted by such devices being produced at the manufacturing stage. Meanwhile, advertising is becoming increasingly integrated into the flow of media content promoting rampant consumerism and extolling the pleasures of consumption as a balm to any environmental impact.

In Chap. 5, Xin Tong explores the other end of the chain of production and disposal, detailing how the dumping of electronic waste in sites on the periphery of Beijing has provided the basis for a thriving microeconomy of scavenging and salvage, generating tensions with plans for urban redevelopment and forcing the waste industry further and further out of the city as land is reclaimed for housing and business.

The two other chapters in this section examine the increasingly central part played by control over data in organising corporate responses to emerging challenges and opportunities.

In Chap. 6, Jo Bates details the increasing corporate capture of data and analysis on weather patterns and the advantages this bestows in devising responses outside the domain of pubic intervention and action. In Chap. 7, following on from his 2014 book, *To the Cloud: Big Data in a Turbulent World*, Vincent Mosco details how command over comprehensive stores of 'big data' on every aspect of daily life coupled with the ability to analyse it is combining with cloud computing and the Internet of Things to build the essential foundations for a new communications system, the Next Internet. While the development of the present Internet was nothing short of game-changing, the Next Internet is already proving even more disruptive. The cloud has become a data analytics factory rather than a simply a storage utility, using the mostly qualitative information available from digital giants like Facebook to find patterns in sample sizes numbering billions and churning out analytics for the use of marketers, accountants and governments. Meanwhile, the Internet of Things produces everyday objects with the ability to monitor their own performance and report to a central system. This new internet is monopolised by a handful of American corporations with concentrated ownership and control structures, sharpening tensions between their pursuit of corporate expansion, escalating demands on resource and energy use and the consequences for climate change.

### Section Three: Corporate Captures: PR Strategies and Promotional Gambits

This section examines the strategies corporations employ in their attempts to influence public debate and policy on climate change.

In Chap. 8, Kim Sheehan, a past president of the American Academy of Advertising who has consulted for numerous consumer product companies, including People magazine and Laura Ashley, and now at the University of Oregon, explains her work with the Greenwashing Index, monitoring corporate public relations strategies designed to present a green image.

In Chap. 9, David McKnight and Mitchell Hobbs detail the public relations strategies used by the mining and energy lobby in Australia to defeat climate change policy initiatives which they see as running counter to their corporate interests. They present three case studies where these strategies were implemented to maximum effect: the Labour government's attempt to implement a carbon emissions trading scheme, the 'charm offensive' mounted by the mining and energy sector to cultivate strategic alliances and allies to prevent future legislative challenges, and the Labor government's second attempt to introduce a carbon tax and the mining industry's powerful backlash. These case studies demonstrate very clearly the ways the mining and energy sector employ varying combinations of 'reward and punishment' that can make it difficult for governments to enact legislation to reduce carbon emissions.

In Chap. 10, Patrick McCurdy explores the struggles around the contentious exploitation of Alberta's bitumen sands focussing on the possibilities and problems presented by the mobilisation of celebrity endorsement of protests. Taking four prominent celebrities–actress Neve Campbell, singer Neil Young, film director James Cameron and actor Leonardo DiCaprio–who have lent support to the tar sands protests by travelling to the disputed sites of extraction and physically 'bearing witness', he demonstrates how easily an inadvertent slip can undermine the credibility of the cause. Of the four celebrities Leonardo DiCaprio had the strongest track record of concern and activism on environmental issues, having established a foundation in 1998 promoting projects to protect endangered habitats and ecologies. But a simple error, mistaking a recurring local weather condition for a more general signal of climate instability, was immediately seized on by media hostile to the protests and used to discredit him and the protest.

In Chap. 11, Benedetta Brevini and Terry Woronov return to Australia, a country with one of the highest rates of emission per head of population in the world, and detail the adoption of 'post-truth' politics in public discourses employed to promote the opening of a huge new open cast coal mine, the Adani Carmichael, in Queensland. Defined as 'relating to or denoting circumstances in which objective facts are less influential in shaping public opinion than appeals to emotion and personal belief' (Oxford Dictionary 2016), post-truth politics has recently been identified as a hallmark of the current era in which forms of political communication and spin that favour feelings and emotions over policy are spreading globally. To explore some of the ways that post truth politics and 'truthiness' operate this chapter explores the way in which politicians and the media in Australia have dubbed the construction of the Adan mine as not only economically necessary but as central to the country's sense of itself and its future.

### Section Four: Communication and Campaigning: Oppositions and Refusals

The chapters in the final section explore the strategies employed by protest movements and campaigns in their struggles for visibility and legitimacy, and examine how journalism can contribute to raising awareness and prompting action.

In Chap. 12, Jodi Dean argues that dominant constructions of "the immensity of calamity of the changing climate-with attendant desertification, ocean acidification, and species loss seemingly forces us into seeing all or nothing", and are creating a pervasive climate of anxiety that is fuelling resignation rather than resistance. Countering this, she explains, requires us to move beyond the totalising construction of the Anthropocene and view contemporary conditions from the side, anamorphically, in search of points that allow us entry into other ways of looking, ways that move us towards collective action and strategic engagement. One powerful route to securing these openings, she argues, is through public art and performance that dramatizes issues in immediately accessible and memorable images. She offers Liberate Tate's performances protesting the museum's acceptance of sponsorship by British Petroleum as an example.

At the same time, given its continuing centrality as the primary source of popular information and interpretation, it is also necessary to think through possible ways of mobilising the news system in the service of increased popular understanding of the causes of climate crisis and how they might be addressed.

In Chap. 13, Robert Hackett and Shane Gunster provide a comprehensive overview of the options available for cultivating journalism that engages more fully and proactively with the challenges of climate crisis. They explore three possibilities: the potential for integrating environmental reporting into journalistic paradigms organised around commitments to civic and peace journalism; the possibility of environmental movements collaborating with supportive journalists to reframe climate politics; and the prospects for vibrant and sustainable alternative media.

In Chap. 14, Alan Rusbridger, former editor of *The Guardian* UK, discusses one major contemporary journalistic initiative, the institution's ground-breaking 'Keep it in the Ground' campaign. Born from a desire to take action commensurate with the problem, it generated a complex public debate and resulted in several major investors, including the Gates Foundation, divesting from leading fossil fuel companies.

In Chap. 15, Greenpeace CEO David Ritter discusses the problems inherent in mainstream climate reporting and the global promotion of consumerism as the key to personal fulfilment, highlighting the relationship between climate change and capitalist reliance on carbon-based sources of energy. The Australian media sector presents a particularly entrenched opposition to calls for reform with dominant players in the commercial news system, led by Rupert Murdoch's News Corp., persistently promoting the illusion that the dominance of man-made climate change remains contestable and placing science on a level footing with opinion under the banner of 'balance'. As an alternative, Greenpeace has frequently utilised online media to spur grassroots activism and apply pressure to large corporations, to great effect.

In Chap. 16, Blair Palese describes her work with 350.org, foregrounding the success of globally-run divestment campaigns over the last 2 years in building a movement to stop new fossil fuels and rapidly reduce greenhouse emissions by countering the power of the fossil fuel industry with the power of people taking collective action,

# PART I

# Communication and Carbon Capitalism:
# Contested Futures

**Fig. I.1** "Break Free" Protest in Thailand (photo courtesy of Greenpeace)

# CHAPTER 2

# An Interview with Michael E. Mann: Fighting for Science Against Climate Change Deniers' Propaganda

## Michael E. Mann and Benedetta Brevini

In 1998, Michael E. Mann developed the so-called 'hockey stick' graph, which revealed sharply higher global temperatures after 1900, to fight against climate change denialism. His most recent work, *The Madhouse Effect*, sees him teaming up with political cartoonist Tom Toles to satirise the twisted logic of denialists.

### CLIMATE CHANGE—THE CURRENT SITUATION

**Benedetta Brevini: The third page of the latest COP 21 Paris agreement acknowledges that the new $CO_2$ target won't keep the global temperature rise below 2 °C, the level that was once set as the critical safe limit. What can be done? Is there still a chance for us to do enough to limit the impacts of climate change? And why was the 2 °C so important?**

M.E. Mann (✉)
Pennsylvania State University, State College, USA
e-mail: mann@psu.edu

B. Brevini
University of Sydney, Sydney, Australia
e-mail: benedetta.brevini@sydney.edu.au

© The Author(s) 2017
B. Brevini and G. Murdock (eds.), *Carbon Capitalism and Communication*,
Palgrave Studies in Media and Environmental Communication,
DOI 10.1007/978-3-319-57876-7_2

23

MM: It's true. And there's been a fair amount of coverage about what the pledges actually buy us in terms of curtailing further warming. And you can tally up the net effect of all the pledges—it gets us about half way from business as usual, which would be 5° by the end of the century. It gets us half way to two degrees—it gets us in the middle, around 3.5°. It doesn't get us down to that two degree mark, which is what many scientists say is the level at which we experience even more dangerous impacts of climate change. But it gets us on the path. The idea is that Paris alone isn't going to solve the problem, but it creates a framework that can be built on further with further reductions at the next conference—major conference of the parties, which hopefully can get us below 2°.

Ultimately, any amount of additional warming is bad. So there's no really fixed level that sort of divides safe and dangerous. It's really more an ever-steeper downward slope rather than a cliff. And further we head down that slope, the further we go down that highway, the worse things get. And we want to take the soonest, earliest exit that we possibly can off that highway. So 2 °C warming relative to pre-industrial is the result of somewhat subjective assessments—when you look across the various sectors that climate change impacts—food, water, health, loss of coastal property, the economy, a whole host of metrics of climate change impacts —and you look at the various studies that have been done estimating how those impacts depend on warming, you find that above 2 °C is where all the impacts really start to look negative. At less warming there's actually the possibility that some impacts are minimal or even slightly positive, but once you get above 2 °C warming, that's where pretty much all of the assessed impacts start to look negative and so you're looking for some reasonable line in the sand to draw where we can say we really see the worst impacts of climate change. Two degrees Celsius is pretty reasonable.

**BB: I understand that. Reading 2036 and 2038 as the deadline for us seems dangerously close.**

MM: Yeah I know, that's right. For us to continue as business as usual with burning of fossil fuels—we pass that two degree threshold very quickly. So without any action we will cross that threshold in a matter of a couple of decades or less.

**BB: In your new book you address the problems of geoengineering as a proposed solution to the climate crisis. What are the issues with this, as you see it?**

MM: The title of our book is *Geoengineering,* or *"What Could Possibly Go Wrong?"* And it really raises the issue of unintended consequences. Experiments that have been done show that iron fertilisation doesn't work very well. That it causes a more rapid cycling of carbon through the atmosphere and the upper ocean, it doesn't bury it permanently in the deep ocean, which is what you need if you're going to take the carbon out of the system. Moreover, the iron fertilisation of the ocean appears to preferentially favour some of the more dangerous algae—like the algae that cause red tides, so it's really an excellent example of how we can end up doing far more damage than if we had not engaged in those interventions at all. So I'm very wary of the vast majority of geoengineering schemes.

There's one that's relatively safe—it's called direct air capture and basically it's trying to suck the $CO_2$ back out of the atmosphere, so you're not really tampering with the climate system in a way that these other schemes are. But it turns out it's really expensive to do that energetically and economically and so probably the only situation in which it might make sense would be if we find ourselves in a situation where we're going to go past one of those dangerous limits and there's nothing we can do—it's too late. Then some argue we need to look for a so-called 'stop gap'—some immediate intervention that we can pull out of a hat and maybe something like direct air capture could be that. But these other geoengineering schemes could lead to far more dangerous impacts on the climate and on our environment and they're more likely to do harm than do good. So my view is that scientists have the same ethical responsibility as doctors—first we should do no harm, that should be our pledge—and geoengineering violates that.

## SCIENTISTS AS PUBLIC COMMUNICATORS

**BB: What should be the role of scientists in society?**

MM: I think it's important to have individuals within scientific communities who are committed to communicating science and its applications. That doesn't mean that all scientists should talk to the media. I know quite a few who would probably never talk to the media. But we do need to provide incentives and support at an institutional level for scientists who want to engage in what I consider to be a very noble undertaking of communicating science to the public. If they don't do that, if there aren't scientists who are willing to play that role, we create a vacuum that becomes filled by the forces of disinformation and denial. And so it's really incumbent upon us to do that. But with that being said, not all scientists should communicate to the media or engage in outreach in general to the public—because a lot of scientists are sort of at their best when they're communicating with their colleagues and they know that if you start using scientific jargon and shorthand in your communications to the public, that is not very effective. And so I think those scientists who do communicate need to learn and understand and train in the rules of effective communication, not dumbing the science down but communicating it clearly, in non-technical terms, without giving abbreviations and jargon.

**BB: What can the scientist community do to communicate more effectively? Could you comment on your own experience of operating your blog?**

MM: I think social media is a very valuable tool for outreach and communication, but there's no one tool in my mind that serves all roles. Twitter only has 140 characters. You can link to an article or something else that provides more context but it's a very fast-paced, on-the-cuff, real-time means of communicating in short soundbites. There is a need for more context. There is a need for pieces that provide far more background, for more content, far more nuance—and you can do that with a blog or by writing commentaries for various online media outlets. I do a fair amount of that—I just had something in the *Guardian* the other day. And there are so many other ways we can communicate: giving public lectures, writing books—as we both have done—trying to explain the issues to the public. And all of these means of communication are complementary. To me they're a part of a larger portfolio. A portfolio of communication. You need a variety of tools in that portfolio to be able to serve all the various roles.

I don't think science journals in general are accessible to most people. Even the generals that try to do that, like *Nature* and *Science*, where at least the first paragraph is supposed to be understandable to a lay audience—or the IPCC reports, the reports of the Intergovernmental Panel on Climate Change. The problem is when scientists think they're being accessible, they're not. They're just being a little less technical than they would normally be. What a scientist views as non-technical and jargonless communication is very different from what we mean in the world of actual communication, when we say non-technical.

But it's important to publish peer-reviewed articles and peer-reviewed science is ultimately what supports much of one's communication efforts—

which is to say that when you're talking about a particular topic, what's the relationship between the extreme heat we've seen in Sydney and this winter and climate change, we can draw upon this peer-reviewed literature. It's there to provide support for the points that you might make, and moreover, if you want to have credibility as a science communicator that comes in part from having your peers respect you. Having them be familiar with your peer-reviewed work. And that process of publishing peer-reviewed literature and doing science is very important in grounding you. Keeping one foot in the world of scientific research I think makes you a better communicator, in part because it helps to ensure you're familiar with the cutting edge of where the science is. And if you're not immersed in the literature, you're not going to be.

## ATTACKS ON SCIENCE

**BB: In 2009 your emails were hacked and used by climate change denialists and mining/oil lobbies to discredit climate science. Several years later, no wrongdoing was found on the part of the scientists.**

MM: You have to recognise that the attacks happened in the lead up to the Copenhagen summit in 2009. They were designed simply to hijack the discussion at Copenhagen. The investigations—eight, nine, ten of them— that found that there were no improprieties revealed in the stolen emails played out over several years and in the meantime climate change deniers were able to exploit the scandal for all its worth.

**BB: And we lost almost decades.**

MM: We lost almost a decade and we continue. It's an attack that can be used over and over again in part, because we have a media that in many cases isn't doing its job and just plays along with the whole false balance. We'll just put it out there—put both sides out there and that doesn't serve the public good.

## DEALING WITH THE MEDIA

**BB: That's interesting. I tend to blame the commercial media for their lack of understanding of environmental issues, their lack of expertise. Lack of funding sometimes ... so what do you think, in general, of the media coverage?**

MM: It's varied. I think there are great media outlets. I've had good experiences here in Australia with the ABC Radio show that I did; the Sydney Morning Herald, a wonderful newspaper with Peter Hannam—a really great guy. There's another person at Sydney Morning Herald—a science person I've talked to before. So they do a really good job. I think the *New York Times* in the US has been doing a pretty good job in covering climate issues. MSNBC, in terms of our cable networks, has done a really good job. CNN has not done a really good job. FOX News is actively promoting misinformation and disinformation, but yeah, you can point to sources and journalists who are doing a really good job. I think the problem is, with the changes in the media environment, there are fewer and fewer positions for those sorts of journalists. Fewer resources—they're understaffed. They don't have the resources that they used to have to do really hard-hitting investigative journalism. That takes resources to be able to do that.

I think we've lost something in the fragmentation of our media. It has made it really difficult for the scientific community to clearly get its message out because it's so fragmented, and you have a variety of media outlets with varying levels of facilitating and accurately reporting science-themed stories. So it's a tough environment and it's, in my view, some of these problems that have led to the fact that technical issues that are contentious, like climate change, too often get treated with false balance. Because you know the journalist, the reporter, doesn't have the resources to investigate who's right and who's wrong—to fact check, to do the investigative work that's necessary ... they often end up resorting to sort of the default, which is there are two sides—and we'll just present these two sides.

**BB: Do you have solution for that? Do you think there is a solution—to stop the media from thinking like that?**

MM: Well Donald Trump has a solution. He wants to imprison all journalists except FOX News. I think that it's difficult because of the corporate media environment. I personally know of many cases where a journalist wrote an article, and I spoke with the journalist and had a sense that they were going to write a really good article. The article appears, it's got some problems, and there's some fake balance—and they throw in the quotes of some industry group and in many cases what you learn is that wasn't the journalist. That was their editor.

**BB: You always have to please the editor.**

MM: And editors are a part of management. So journalists are sort of workers and the editors are management, it's sort of the same workers and management divide. You don't want to blame the workers because they're doing what workers are meant to be doing. In many cases the blame goes to the management.

## Contributing to Policy Making

**BB: It's a not a fair marketplace. In general, when you see that policymakers don't take your advice and don't take into consideration your findings and your studies. How does it make you feel?**

MM: Again, it's varied. There are a lot of really good policymakers that I've advised. Jerry Brown, the Governor of California, I've been an advisor to him and he's doing wonderful things with renewable energy and climate change: putting a price on carbon, helping the former coalition of Western states to price carbon, incentivising renewable energy. And he's taken on Donald Trump and the rhetoric that Trump has been using when it comes to climate change. Jerry Brown is sort of a pitbull; he's fought back against efforts of Congressional Republicans to misrepresent the science. Sheldon Whitehouse, a senator of Rhode Island, is sort of again a pitbull on the senate floor—every week giving a speech about climate change; calling out climate change denialism. Whether or not anyone's willing to listen to him, he's down there on the Senate floor. So there are still some really good politicians in the US and elsewhere who understand the threat that climate change represents and want to act and do something about it. There are even a few on the Republican side of the aisle who quietly support action but are afraid of putting out and saying so as they'll be vilified by the very same fossil fuel interests we've been talking about.

At the other extreme, you have politicians who are just mouthpieces for the fossil fuel industry. Their campaigns were funded by the fossil fuel industry; they have close personal ties, some of them even benefit directly —financially—from the fossil fuel industry. Just about every individual who has been appointed to the Trump's administration and cabinet at this point is a climate change denier and has close ties to the fossil fuel industry. Like Rex Tillerson, the Secretary of State, the CEO of Exxon Mobile.

## CONTESTED FUTURES

**BB: Before Trump, just after the climate summit, there was a bit of optimism. So how do you see the future now? Do you think this wave of optimism has been stopped by Trump?**

MM: I think that there are a number of things going on. Firstly there's progress at the global level and that's really important: regardless of what Trump does, you can't stop the momentum that now exists for progress on climate change—for transitioning away from fossil fuels towards renewable energy. The rest of the world is going in that direction and there's nothing that Trump can do to stop that. What he can do is make the challenge a little harder. Delay. And in so doing make the challenge even greater to limit warming below dangerous levels. The amount of carbon that will put us over the edge if the US pulls out of the Paris accord, for example, will make it even more difficult for us to stay within our carbon budget. That having been said there's a lot of hard work that is happening still at the municipal level, the state level, business that are committed to lowering their carbon footprint.

**BB: There are the campaigns as well ....**

MM: The campaigns—absolutely. Getting major corporations and politicians to divest holdings from the fossil fuel industry. So there are all of these good things that are happening and, to me, they signify that we will ultimately prevail in this battle. The question is will we allow Trump and other bad actors to slow progress down enough that a lot more bad impacts are observed. That we commit to worse climate change impacts. That's my fear. Ultimately it won't prevent us from solving the problem, but it will ensure that we commit to more harm and damage and suffering than we otherwise would have.

CHAPTER 3

# An Interview with Naomi Klein:
# Capitalism Versus the Climate

*Naomi Klein and Christopher Wright*

In her latest book, *This Changes Everything* (2014), Canadian writer and activist Naomi Klein argues that unrestrained capitalism is at the root of the climate crisis and that the global response to climate change has, thus far, been shaped by wealth and power.

**Christopher Wright: Naomi Klein, your book, *This Changes Everything* has proven hugely popular in tackling the vexed issue of climate change. Unlike a lot of the existing debate on the topic, you're quite explicit that climate change is not so much an environmental issue but one that derives fundamentally from our economic system of neoliberal capitalism. What particular events led you to this realisation?**

NK: Well I guess the key event for me was Hurricane Katrina ten years ago. At the time I was working on my previous book on *The Shock Doctrine* and that book begins and ends with Hurricane Katrina. I reported quite extensively on the storm and its aftermath as an example of what I call disaster capitalism. The disaster in New Orleans was really about a collision between climate change (between heavy weather linked to climate change)

N. Klein
Toronto, Canada

C. Wright (✉)
The University of Sydney, Sydney, Australia
e-mail: christopher.wright@sydney.edu.au

© The Author(s) 2017
B. Brevini and G. Murdock (eds.), *Carbon Capitalism and Communication*,
Palgrave Studies in Media and Environmental Communication,
DOI 10.1007/978-3-319-57876-7_3

and the legacy of neoliberal capitalism (and the reality of neoliberal capitalism), as well as American racism, and those three forces intersected in the most toxic way imaginable. Both during the storm and in the years since, in the way that storm was capitalised upon by elites in the United States.

So what I saw in New Orleans was just how really antithetical a political ideology is that does not believe in the state. How antithetical that ideology is to what needs to happen in the face of climate change. At the time, Paul Krugman called it the 'can't do government' and it couldn't do anything in the face of this disaster. So FEMA, the agency that should have been evacuating people and dealing with this disaster couldn't find New Orleans for five days, and Americans were completely shocked across the political spectrum. It was a totally hollowed out state.

But then it was also incapable, and this is what's more important, in learning the lessons of the disaster. The lesson of that disaster is climate change is real, we need to get off fossil fuels and we need to invest in the public sphere both to deal with the impacts of climate change and to stop making it worse. We need to change our energy system, we need to have public transit, we need to change the way we move ourselves around. None of that happened. In fact New Orleans has become a laboratory for privatisation of various kinds. It's a much more unequal city than it was before the storm.

So, this is what climate change looks like in hyper-capitalism and it looks like New Orleans and it's not a pretty sight.

**CW: You argue that the current climate crisis is a product of what you term 'bad timing'. Could you elaborate on that?**

NK: Scientists have understood the connection between greenhouse gases and warming for a long time but the issue had its tipping point moment in the late 1980s. That was the moment when we all lost all plausible deniability; 1988 was the year when governments first had an intergovernmental meeting to talk about the need for emission reductions. That was really the turning point year. It was also the year that the Intergovernmental Panel on Climate Change was formed.

So I think one thing you can only see with hindsight is what an epic case of bad timing it was. What else was happening in 1988? Well, Canada and the US signed their first free trade deal that became the prototype for NAFTA and then all these other trade deals that have proliferated around the world. It's the year before the Berlin Wall collapsed. It's right when Francis Fukuyama declared history over, and this single ideological project was then spread throughout the world. Privatisation, deregulation cuts to government spending. Sometimes it's called neoliberalism, sometimes it's called market fundamentalism, sometimes it's call the Washington Consensus, the French call it *pensée unique,* there's no consensus about what to call it. We know what it is because we're all living it. That is an epic case of bad timing because at its core it is a war on the collective. It's a war on the idea of collective action, it's a liberation project for capital, that's what it is and it's been a very successful one.

It's not compatible with a crisis like climate change because climate change is the essence of a collective crisis that requires that we act together within our countries, between our countries. A winner-take-all ideology does not compute with a crisis like this that requires that we see how we are interdependent.

But there's more to it than that. We were systematically selling off exactly the parts of our economies that we most needed to control if we were going to take climate change seriously. Our rail systems, our energy grids, our water, this is what neoliberalism did. That makes what we need to do so much harder. Just because something is publicly controlled doesn't mean that it's good, doesn't mean that it's environmentally conscious, but if a public has the ability to have a say over their energy grid, they can say they want it to change and that's what we really haven't been able to do.

That's why I don't think it's a coincidence that the countries that have taken some of the boldest moves in the face of the climate crisis are ones that are most socially democratic. It's not a coincidence that the Scandinavian countries have some of the most enlightened climate policies (and put a big asterisk next to Norway). Or that Germany, which never fully embraced neo-liberalism for historical reasons, though they prescribe it with brutality on the rest of Europe, they know at home that it's very dangerous for them to get rid of their safety net. Germany, because of that, has been willing to introduce a very ambitious feed-in tariff programme that has transformed their energy grid very rapidly. So that's the conflict.

**CW: You make a very strong argument that neo-liberal economics is driving humanity's greenhouse gas emissions. However I was also struck by your point that these are modern expressions of an older logic particularly the view that nature is something we can 'bend to our will'. To what extent then is the problem, neo-liberal capitalism, or our reliance on fossil fuels, or is it something deeper in the relationship between humanity and the natural world?**

NK: We've been talking about this collision between neoliberalism and climate change and it does go deeper than that. Now because we've waited so long we need to be cutting our emissions so rapidly that it isn't in any way compatible with the growth based economic system. The Tyndall Centre says that wealthy countries like Australia or where I live, Canada, we need to be cutting our emissions by eight to 10% a year. There isn't an economist in the world that can tell you how you do that within a growth based economic system which is why the book is not called 'neoliberalism versus the climate'. It's called 'capitalism versus the climate' because that growth imperative is at the heart of our system.

Your question about whether it's even deeper than capitalism and whether it's something about humanity, it's a complicated question. I think it is something deeper than capitalism and we know that industrial socialist economies have been equally violent towards the planet, whether it's Mao's 'war on nature', that's what it was called, proudly, the war against nature. How's that for a slogan? We know that the only time there's been a sustained drop in greenhouse gas emissions, has occurred at two points. One when the Soviet Union collapsed in the 1990s and one when capitalism collapsed in 2008, they both led to severe drops.

So what we know is that the Earth responds well to both of these systems crashing. Now we don't want to crash, we want a great transition to another economic system.

The part that I disagree with is the idea that this is about humanity. It's not all of humanity that is responsible for this. In fact it's quite a small minority of humanity, so I think that really at the core of what we're dealing with is an idea that took hold in the 1600s in a very specific place, England, and spread to other parts of Europe.

That was the idea that the Earth is a machine that all could be known and the key philosophers of this were Francis Bacon and René Descartes who said man could be the masters and possessors of nature. This is still a minority view if we look at the whole globe. Most people on earth actually approach the natural world with reverence, humility and a healthy dose of fear. I really think you have to be careful of throwing words like 'humanity' around.

The other thing that's complicated about it is that this idea emerged at the same time as industrial capitalism was emerging. So you really can't pry it apart from the emergence of capitalism. What came first, right? The fact that the Industrial Revolution was kicking off or that René Descartes had that idea. What we do know is that it took the commercial steam engine, the marketing of the commercial steam engine in the late 1700s. To take that idea which was just a theory, this idea that we could separate ourselves from nature and dominate it and know it and no longer be at its mercy, and turn that into an apparent reality.

So it was the combination of an idea and a technology that allowed us to really convince ourselves, this small subset of humanity, that nature was a thing and that we could be the boss.

I think the reason why climate change is so threatening is that it is a rebuke, a fundamental rebuke of that idea and it is saying to us that we were never the boss. This was an illusion and in the book I quote Robert Manne who says this very eloquently. This is a civilisational crisis, this is a narrative crisis because all the time we were liberating ourselves from nature because we were able to sail our ships whenever we wanted. We didn't have to wait for winds to fill the sails because we were the boss. We could build our factories wherever we wanted. We didn't have to look for rushing water as they did before with the industrial burning of coal. All that time we were burning carbon and it was accumulating in the atmosphere. The response didn't come right away.

So we had the illusion of a one way relationship but it was a fantasy. That's the thing about climate change that makes it more than just an issue. That makes it this narrative crisis, this civilisational crisis because now all that carbon that has been accumulating over hundreds of years is creating a response that is making us feel very weak indeed. We are up against forces that show us that we were never the bosses that we imagined ourselves to be. I think that it's a crisis of story, it's a crisis of relationship but I would be very careful about attributing it to humanity. Because it comes from a place and not all humans believed it and still don't.

**CW: Of course there are many in business and government who promote a vision of green capitalism. That the answer is to use market forces to price carbon, to drive decarbonisation and decouple growth from its material impacts. What are your thoughts on such arguments? Is green capitalism and green growth a myth?**

NK: Is it a myth? I think it needs to be pried apart in that there are green pockets within capitalism and it is possible to have some marginal growth while lowering emissions, I think that's true. Certainly transitioning away from fossil fuels, if you do make the sorts of investments that we're talking about, changing an energy grid, changing a transit system, this is going to create huge numbers of jobs and there is going to be growth. The problem is there also has to be contraction at the same time because we need to be lowering our emissions.

So people who get carried away with the green growth idea, they know how to add but they're not so good at the subtraction part and I quote Kevin Anderson who is a really important climate scientist and emission reduction specialist at the Tyndall Centre, he's the deputy director and used to be director. I quote him a lot in the book and this quote isn't in the book because he said it since the book came out but he said you have to make a distinction between going more slowly down the wrong road and getting on the right road. A lot of this green growth stuff is about going more slowly down the wrong road.

**CW: In the latter half of your book you focus on social movements that are emerging in response to the climate crisis particularly the phenomenon you term 'Blockadia'. For me this was quite an optimistic message. Who are the natural leaders of the climate justice movement and how do we make sure their voices are heard, given that many are acting at a very grass roots level?**

NK: The people who are leading this movement are the people who are most directly impacted by extraction and other forms of fossil fuel infrastructure whether it is pipelines crossing their lands or coal export terminals impacting their fishing grounds. So overwhelmingly it's people who still live off the land which means that they're overwhelmingly indigenous people or farmers and fishing people. They are building this movement with incredible speed and I think what's exciting is the intersection of this place based movement that is really driven by love of place.

One of my favourite quotes from the book is from a woman named Alexis Bonogofsky who is a goat rancher in Montana and she says this is what the coal companies will never understand, that our movement isn't driven by hatred of them. It's not driven by hatred of the coal companies, it's driven by love, love will save this place. I think that just from everything I've seen that is absolutely the driving force. It's love of land, love of one's kids and a duty to protect for future generations a different relationship to the land that is non-extractive.

It isn't even about stewardship in the sense of just taking care of the land so that it takes care of us. It's more about an ethos of caring for the land and caring for one another.

I mean the exciting part is that intersection between these very local struggles and technologies that allow these different various front lines of

Blockadia to learn about one another and to find each other and to feel themselves part of a truly global movement. So when there was a huge climate March in New York City last September when there was a UN climate conference, what was beautiful about that March wasn't just that it was huge, and it was huge. It was the largest climate March in history, 400,000 people. It was that it was this collection of impacted people led by indigenous people at the front, a huge anti-fracking movement that has since succeeded in banning fracking in all of New York State. A big contingent from the South Bronx who had signs about the very high asthma levels that their kids were suffering but also demanding green jobs for their communities and services that would make their lives tangibly better.

I think what's significant about this is not just that it's a movement that looks like our countries. Our movement should look like our countries as opposed to a tiny subset of our countries which is what they do often look like. It's also that I think that this kind of movement where people have so much on the line for better and worse, has the potential to bring much needed jobs and services to really neglected communities and also better health. These are in many cases really life and death struggles.

People in a movement like that fight differently. They fight really hard because they can't afford to lose and I think too often what the climate movement has suffered from this kind of thing that this is a movement for people who don't have anything better to care about or something like that. There's this thing of it's a luxury concern for people who are very privileged. I think what's really changing is the emergence of the climate justice movement taking centre stage which brings together those daily economic concerns through justice, jobs, services, health and the need for climate action. That kind of movement I think has a much better chance of winning against players in the fossil fuel companies who are themselves fighting for their lives. They are fighting for their lives because if we win their business model is cooked. So I think that that's very exciting.

Your question about how do we hear more from them, I do think that's a very important question for the green movement which still is scandalously white and middle class. The leadership is too male in a movement that at the grass roots is overwhelming led by women. It's a huge problem and I think it's a problem on a lot of different levels. This is a movement that is asking governments to change very rapidly and I think that the onus is really on the climate movement, the environmental movement to model change.

If we are going to ask our government to change so quickly then I think we also have to look at our own house and say why have we been having these same discussions for 30 years about why we're too white or too middle class, and doing so little about it in terms of sharing resources and sharing the spotlight? If we can't change then how do we expect to have any credibility asking our governments to change?

**CW: Politicians here in Australia are fond of pointing out we've barely 1% of world emissions so Australia can't really claim a leading role in climate action. What do you make of this argument and what do you think Australia's proper role in climate action should be?**

NK: Look, Australia has the highest per capita emissions anywhere in the developed world. Australia wants to open the largest coal mine anywhere in the world. Australia has some of the dirtiest coal fired power plants in the world. Australia is not irrelevant and this is just an excuse that we've been hearing for too long and this is why we have international negotiations by the way. We have international negotiations because nobody can do this on their own. We have to come together and we have to do it in a way that we all agree to.

There is a UN climate convention that agrees to the principle of common but differentiated responsibilities, which means that the solutions should reflect the fact that Australia had a 200 year head start on burning coal. So that means that there's a *greater* responsibility to lead and when countries like Australia, Canada and the United States make a bold commitments that makes it harder for countries like China and India to resist. It strengthens the movements in countries like China and India that are wanting to leap frog over fossil fuels and those movements are increasing strong. So I think that's a pretty ridiculous argument.

# PART II

# Toxic Technologies: Media Machines and Ecological Crisis

**Fig. II.1**  Little girl at waste dump site (photo courtesy of Greenpeace)

# CHAPTER 4

# Digital Technology and the Environment: Challenges for Green Citizenship and Environmental Organizations

*Richard Maxwell and Toby Miller*

This chapter focuses on the environmental impact of media and information and communication technologies (ICTs) and the challenges that green citizens and environmental groups face in greening their use of ICTs.

For most of us, media technologies hardly seem like the kind of human creation that could cause any significant environmental harm. Media certainly appear to be a clean industry, especially electronic and digital media. Books too seem to provide low wattage entertainment and enlightenment, while movies and TV shows emit no apparent exhaust. Smartphones glow with an aura of clean energy as they link us to networks that have come to epitomize the post-industrial era, a time far removed from the realities of smokestack capitalism. With this benign view of media technology, it's no

R. Maxwell (✉)
Queens College, New York, USA
e-mail: rmax@nyc.rr.com

T. Miller
University of Loughborough, Loughborough, UK
e-mail: tobym69@icloud.com

© The Author(s) 2017
B. Brevini and G. Murdock (eds.), *Carbon Capitalism and Communication*,
Palgrave Studies in Media and Environmental Communication,
DOI 10.1007/978-3-319-57876-7_4

wonder that media scholars, commentators and school curricula have virtually nothing to say about media technology's impact on the environment.

In response, we began a research project over a decade ago that we hoped would disabuse media and communication experts of the idea that media, information and communication technologies are environmentally neutral. We have drawn on this research for this chapter, primarily from our book, *Greening the Media*, and a column of the same name published monthly at http://www.PsychologyToday.com. Our aim continues to be to spark new ways of thinking about media history, media industries and the connections between media usage, climate change and our health. Our specific focus here is on challenges for green citizens and their expectations of environmental organizations engaged in some form of activism, advocacy, or policy-oriented work.

## The Ecological Crisis and the Technological Challenges

Readers of the collection of essays will be familiar with the basic constituents of the ecological crisis:

- global warming (climate change) caused by overproduction of carbon dioxide ($CO_2$);
- pollution, including industrial dumping, unabated and unregulated;
- rapidly diminishing biodiversity, the "sixth great extinction" and the precipitous decline of biosphere integrity (loss of land and sea habitats).

Climate and environmental scientists have different ways of explaining the ecological crisis, but they agree on one point: that since the industrial revolution people have dangerously tipped the balance between what the Earth can give to support human activity and what the Earth can safely re-absorb from those activities. This balancing function has been called the "scientific prerequisites for ecological sustainability" or, more simply, our "planetary boundaries" (Rockström et al. 2009a, b; Schauer 2003).

Researchers who developed the model of planetary boundaries have shown that human industriousness has crossed four of the nine boundaries in their model,[1] and that the transgression of two of the boundaries that are key to the resilience of life on the planet—climate change and biosphere integrity—"could inadvertently drive the Earth System into a much less hospitable state, damaging efforts to reduce poverty and leading to a deterioration of human wellbeing in many parts of the world, including wealthy countries" (Crutzen and Steffen 2003). We have already witnessed such inhospitable moments in severe weather events that are happening with more force and frequency, while droughts and rising sea levels continue to threaten lives of millions of our fellow inhabitants, human and non-human.

Environmental organizations and activists have employed ICTs to help us better understand these changing conditions while strengthening forms of resistance to, and reforms of, wasteful and toxic practices. Satellites and networked telecommunication combined with innovations in measuring and monitoring techniques have improved how we track atmospheric changes, the health of sea habitats and marine life, and species well-being. There are new processes for recovering and reusing harmful plastics arising from advances in green chemistry (Grossman 2009). There are even mobile applications to enable greener consumption (Maxwell and Miller 2014).

Meanwhile, citizen journalists and non-governmental organisations (NGOs) are collecting evidence of despoliation of the natural environment and publicizing their findings via digital channels[2]; confrontations between Native American water protectors and oil-barons' police armies have been recorded in daily video dispatches from North Dakota to be seen by millions on social media screens[3]; powerful video and photographic evidence of illegal toxic e-waste being picked over and burned for precious contents in far-flung salvage yards has inspired mitigation reforms (*Recycling Today* 2015); artist-activists are mounting exhibits of digital works to think through the traumas of displacement and forced migration caused by the warming of the planet.[4] Workers in the electronics factory zones of China use the very same smartphones they assemble to memorialize their struggles and share songs, slogans and poetry in solidarity with their comrades (Qiu 2016).

While high-tech tools are important for organizing responses to global warming and its pernicious effects—from techno-scientific mitigation projects to environmentalists' advocacy and activist efforts—the social liabilities they carry with them must also be counted among the problems of the ecological crisis. Good green intentions do not exempt progressive uses of digital technology from having negative effects associated with environmental decline: rising energy consumption, associated carbon emissions, toxic working conditions and wasteful consumption habits.

Consider that there are upwards of 14 billion networked devices across the globe, along with a huge private and governmental ICT complex, telecommunications structures and so-called cloud services running millions of servers inside large refrigerated warehouses. This vast interconnected system of digital devices runs on energy grids that produce carbon emissions that rival most major industries—at the same time, these technologies have become vital to all major industries. The manufacture of digital technology relies on raw materials, some highly toxic to biological organisms that are too often mined in unsafe, slave-like working conditions. The factories where high-tech goods are assembled and receive their final polish have been likened to a plantation system, where workers are driven to produce at inhuman speeds to match the insatiable demand for ever new or upgraded devices (Qiu 2016a, b). The manufacture, use and disposal of high-tech goods have real bio-physical effects on workers and ecosystems.

Environmentalists who grapple with this downside of digital technologies face a paradox about how to ensure ecologically sound practices in their organizations. For while the expansion of their digital activities is perceived as inherently good, such virtue does not magically eliminate the bad environmental impact of the ICTs they use. This is similar to the dilemma faced by cultural organizations aspiring to do their part for sustainability (Maxwell 2015a). Of course, the cultural sector is more deeply embedded in capitalistic economic development, with global trade in cultural products increasing from US$559.5 billion in 2010 to US$624 billion in 2011, for example (United Nations 2014). The European Commission defines cultural and creative industries as an economic growth sector, with emphases on education, artists' mobility, regulatory reform and market access and investment.[5] For its part, UNESCO promotes culture as a fourth pillar of sustainable development, an idea that elevates creative industries to equal partnership with stakeholders working for economic growth, social inclusion and environmental balance. The aim is to win a place for "culture-led development" in sustainability debates, not environmental sustainability within the cultural sector (UNESCO 2012).

Likewise, environmentalist organizations seek to lead discussions of development to a greener path, but not necessarily by looking at their own internal sustainability practices. This is exemplified by the policy platforms of major green parties. In Germany, Canada, New Zealand and the UK, the Greens' cultural policy has focused on identity, heritage, institutions, funding, ownership participation and involving artists and cultural workers in sustainable practices and ideas (Bündnis 90/Die Grünen; The Green Party of Canada 2015; Green Party of Aotearoa New Zealand 2014). The US Green Party lacks a cultural policy, except in relation to mainstream media channels, again with an emphasis on democratic principles, access, ownership and so on.[6]

Intervention in cultural practices plays a key a role in green political party efforts to meet sustainability goals, but their programs are primarily focused on resisting commercial culture and the hyper-consumerism that undermines local participatory culture—a laudable goal, but perversely exclusive of any reference to greening operating practices themselves. Such policy platforms only reinforce the idea that the movement for sustainable development applies beyond the realm of cultural production or artistic practice.

Similarly, policy discourses and platforms within environmental organizations fail to acknowledge that their internal practices have social liabilities that directly affect the environment through the material practices of running a pro-environmental operation (Stonebrook 2014). For instance, the National Resources Defense Council (NRDC) says that they "strive to build a cleaner, more sustainable future in our programmatic efforts, operating practices, and financial management". But they do not give details on how they manage their operations in a clean and sustainable manner (National Resources Defense Council 2016). Greenpeace, by contrast, have posted a clear statement on how they use 100% renewable energy to run ICTs in their Dutch facilities, while striving for sustainable energy use throughout their worldwide operations.[7] Most green NGOs lack this kind of explicit statement on internal sustainability activities.

Green citizens look for inspiration and involvement in green organizations, especially ones that practice what they preach. If green organizations' internal policies reflect their external mission, they should publicize this fact with clear governing policies on their internal sustainability efforts and through transparent accounting practices that demonstrate how well they balance the benefits of activist and advocacy work with the environmental costs of operations. Some areas that could be addressed include procurement practices, energy consumption, occupational health, and e-waste.

## PROCUREMENT PRACTICES AND THE SUPPLY CHAIN

Green organizations are well positioned to identify and analyze how their ICT usage is tied into an interconnected supply chain that's spread far and wide across the globe. For example, environmental NGOs, like Good Electronics, have taken the lead in analyzing labour conditions around the world in the ICT sector, reminding us that digital technologies are not post-industrial products despite their appearance of being so clean you could eat off of them.[8]

It would therefore be an important act of consciousness raising were environmental groups to publicize the industrial origins of the ICTs they use. It would help to counter the ideological enchantment with digital technologies—even environmentalists are vulnerable to the fetishism of commodities—that overtakes our perception of ICTs when labour and harsh working conditions are forgotten. So, environmental organizations should be willing to recast their virtuous image with an honest, if disagreeable, accounting of the pernicious effects of the digital technologies they procure. To put the matter of their own procurement practices in the context of the international division of ICT labour would help these organizations and their adherents to identify points of alliance with labour activists and researchers mapping global supply chains, especially in the farthest reaches, and to collectively address problems encountered.

While it requires effort to learn about global and systemic connections between ICTs, supply chains and environmental groups and green citizens, it takes even greater commitment to incorporate that knowledge into reflexive thought and social habits. ICTs are born from a toxic and unsustainable process that begins in mines around the world—the major mining regions are in Africa and Latin America, points of origin for the copper, gold, tin, coltan, lithium and other elements that go into smartphones, tablets and computers. Components assembly and finish work reach around the globe too, but again most of the contract manufacturers that make brand name products are found in Asia and Mexico. Spreading information related to these international realities is one aspect of environmental action; transforming that information into cognitive and behavioural routines is another, though more difficult, process (Maxwell and Miller 2016a).

Green organizations can help initiate and reinforce green routines by participating in and publicizing ongoing research on ICT labour worldwide. This is no simple act of posting a link to what's already known, for there is no agreement on how to describe and map the global workforce involved. Some studies suggest that there are over 200 million people in OECD countries and the Asia-Pacific region alone who work in some kind of ICT job. Add another 200 million for mining and related work, and some additional millions for transport, distribution, retail sales and end of life management, disposal and destruction (Maxwell 2015b). Once involved in the collective effort to improve data collection and analysis, environmental organizations could more readily explain how the lives of workers around the world connect to the deployment of digital technology for environmental action.

This effort would overlap with two political goals in the labour movement, both vital to the improvement of working conditions in the global supply chain. First, it broadens institutional pressure to get accurate information about far-flung and often high-security factories where labour, especially in the assembly stage, is geographically dispersed via a system of international subcontracting.[9] Second, acting to improve research on working conditions would fuse aims of environmental NGOs with union organizers who struggle to build unions in global ICT and electronics industries where union membership density is shockingly low. Weak union representation, probably more than geography, makes it difficult to find reliable and representative information on workers' exposure to toxic materials and other workplace hazards.

## Institutional Energy Consumption

It's surprising that more environmental organizations aren't as forthcoming about their ICT energy consumption as Greenpeace. Supporters might not understand the scale of energy consumption associated with ICTs, so it's important to remind them that like all consumer electronics, digital tools of advocacy and activism also need to be plugged into the electric grid —and that batteries, too, have built-in energy costs in production, usage and disposal.

According to the International Energy Agency, residential electronics alone consume about 15% of the total global residential energy in use. Without any changes to this trend, it is estimated that the residential electricity needed to power this stuff will rise to 30% of global consumption by 2022 and 45% by 2030. This includes about 80 billion dollars' worth of electricity wasted while these network devices are on "standby" (International Energy Agency 2009).

Keep in mind that residential use refers to *operational energy* and not to the energy consumed in the manufacturing stage of information and communication technologies, or in the disposal and recycling stage. Energy used in the manufacturing of laptops, for example, is 64% of the total used in a laptop's life-cycle—and this does not account for the energy used to make chemicals and gases that go into the production of semiconductors or the energy used to dispose or recycle them (Williams 2011).

In addition, enormous amounts of data pass daily through massive networks and data centres—the 'cloud'—now scattered across the globe. Data centres' energy demands rise at a steady pace, with business practices that range from serious plans to reduce reliance on coal-fired energy to widespread examples of waste and thoughtless energy management. At current levels, cloud computing eats up energy at a rate somewhere between what Japan and India consume (Greenpeace 2012). The environmental impact of this networked culture depends on the type of energy production used to power the grids—coal-fired power being the biggest menace.

Recent life-cycle studies tell us that the biggest communications culprits for unsustainable energy consumption are the wireless access providers that link us to the network and cloud, estimating that 90% of the total energy consumed by our networked devices is attributable to access providers—this is not counting the energy used by the devices themselves (Centre for Energy-Efficient Telecommunications 2013).

Green organizations would do their employees, allied activists and supporters a great service by reminding them that smokestacks and pollution accompany electricity needed for high-wattage operations of digital networks, office equipment, video displays and digital living in general. These are not environmentally benign technologies. Environmental organizations should publicize this fact and demonstrate how they address the liabilities of their own institutional energy consumption.

## EMPLOYEE HEALTH

Related to the physical impact of digital technologies' energy consumption is the effect they have on personal health of those working in green organizations (Grossman 2016). Like most contemporary organizations, NGO offices are equipped with an array of ICTs and peripheral devices. Photocopy machines emit clouds of particulate matter that can adversely affect the health of office workers (Elango et al. 2013). Mobile devices have been scientifically linked to biological harm: cases include cancer clusters around cell towers and concerns over effects of long term exposure to radiation emitted by mobile devices. By demonstrating how staff members' occupational health and safety is monitored and cared for, environmental NGOs can provide their supporters with a model for the digital workplace, and also further demystify the clean-industry image of digital technology.

With the advent of manufactured, artificial electromagnetic fields (EMFs) we have altered our electromagnetic environment on a scale that is unprecedented in our evolutionary history. The consequences of this intervention are only beginning to be understood by scientists.[10] But the big brands that market mobile devices know that their radio frequency emissions can cause harm. Researchers have studied the thermal effects of this radiation in laboratory experiments, which measure the range of radiation from a stationary device without taking into account the modulated bursts of radiation that happen every 900 ms when networked devices report to adjacent cell towers. In that moment, whether or not you're using the phone or other gadget, the radiation reaches critical levels.[11]

It's not surprising that environmental organizations tend to provide little help to employees on these issues. There's a widespread assumption among consumers that mobile devices, and wireless transmission more generally, are harmless. Even by conservative estimates, this is questionable. The World Health Organization's International Agency for Research on Cancer advised that the group of frequencies that includes cell phone emissions "is possibly carcinogenic to humans". The US National Cancer Institute urges people to be cautious because additional research is needed to analyze these rapidly changing technologies. The American Academy of Pediatrics called on the US Federal Communications Commission and the Food and Drug Administration to revise EMF standards to account for different peoples' vulnerability to cancer from cell phones, notably pregnant women and children (National Cancer Institute 2012; McInerny 2013).

And yet, it's easy to find published research showing little to zero negative health effects. Many of these studies tend to be industry-funded, a part of a 'doubt industry' that hires 'sceptics' to muddy public thinking with a claim that there are two sides to this story—a common problem in mainstream media reporting on climate change as well (Oreskes and Conway 2010). The main tactic of these merchants of doubt is known as 'war-gaming' the science—attacking any evidence of harm—a practice exemplified by how well it worked for the tobacco corporations for many years (Davis 2013). At least when the tobacco industries lied to us, we could point to a naturally occurring control group of healthy non-smokers. There is no control group in the case of artificial EMF exposure.

Environmental organizations could help by making the precautionary principle a defining feature of workplace safety. This notion basically says that what we don't know about a technology for which scientific consensus is lacking far exceeds what we do know. So, to be on the safe side, we should avoid cavalier certainties about risks associated with mobile technology, especially while ongoing research continually shows greater evidence of harm (Maxwell and Miller 2016a, b). Here again, environmental NGOs can play a demonstrative role by enacting the precautionary principle while, at the same time, identifying, analyzing and publicizing the most up-to-date and independent research.[12]

## Institutional e-Waste

In conjunction with improving workplace safety and implementing transparent reporting on procurement practices and energy consumption, environmental organizations must also be explicit about their end-of-life management of older or disabled technologies. Green citizens expect nothing less.

Behind the projections of rising energy demand is the tremendous growth in sales of electronic equipment. This growth is now mostly made up of sales of mobile devices like tablets, notebooks, and smartphones, of which a quarter is attributable to the US alone. Such growth is the result of the unrelenting marketing of wireless gadgets promoted as tools to keep us constantly connected to each other and the network.

But our love affair with high-wattage high-tech goods is also the leading cause of electronic and electric waste—also known as e-waste—which consumers produce annually at a rate of 20–50 million tons worldwide. Sometime in the last decade or so, e-waste became the fastest growing part of all the stuff we throw away. And this stuff is full of toxins that, if not properly removed, reused or recycled, can poison the land, air, water and the bodies of workers exposed to the chemical contents.

Waste is a problem throughout the life cycle of any electronic device, from water over-used and contaminated in semi-conductor production to discarded solvents and other materials. While such waste is presently designed into high-tech goods, there exist promising alternatives of green chemistry and digital design that could today be put into production (Grossman 2009). But the pernicious business strategy of planned obsolescence still dominates corporate thinking in the high-tech sector.

This destructiveness creates a searing reality for those working in low-tech e-waste dump sites around the world. Wealthy high-tech nations dump 80–85% of their e-waste in Latin America, Eastern Europe, Africa and Asia. Recent estimates from the United Nations suggest that China now receives 70% of all e-waste (Watson 2013). Studies in the European Union show that only a third of all e-waste is identified as such and "separately collected and appropriately treated". Unidentified portions of the rest of this e-waste are probably "going to landfills and to sub-standard treatment sites in or outside the European Union" (European Commission 2016).

Health and ecosystem risks associated with exposure to burned, dismantled and open-pit disposal of e-waste in low-skilled, low-tech salvage yards are well-known. Health risks include brain damage, headaches, vertigo, nausea, birth defects, diseases of the bones, stomach, lungs and other vital organs and disrupted biological development in children. These conditions result from exposure to heavy metals (lead, cadmium, chromium and mercury), burned plastics and poisonous fumes emitted when melting components in search of precious metals (StEP 2016).

To understand ecosystem risks, consider the recent history of Guiyu in Guangdong Province, China. Once a farming area, 80% of local families left farming for recycling, contaminants from recycling saturate the human food chain, and persistent organic pollutants in the soil and water prohibit the safe return of affected agricultural lands to future generations. After worldwide publicity of this disaster, thanks largely to the green NGO, Basel

Action Network, the Chinese authorities restricted illegal dumping and foreign imports of scrap into Guiyu. They also confined all e-waste processing into a huge new industrial park there. The tragedy is that even if people wanted to return the area to some agriculture, the odds are that they'd only produce poisoned crops (*Recycling Today* 2015; *Toxic Leaks* 2016). This story is repeated throughout the Global South (Rams 2015).

Green NGOs must be proactive in their treatment of e-waste. That means explicit guidelines for employees and associated activists and organizations. It means providing transparent reporting as unvarnished as they would with green procurement practices. It means modelling for all green citizens the routines of repair-reuse-recycle that define green institutional behaviour. This not only helps reduce ecologically harmful aspects of digital activism and the environment, it also expands the grounds on which international and intergenerational solidarity can be identified and enacted.

## CONCLUSIONS

More and more consumers are re-evaluating their love of digital technologies and starting to reduce the number of new gadgets they buy and to recycle old electronics as another routine duty of environmental citizenship. Environmental citizenship now informs broader efforts to protect biophysical health from risks associated with toxic substances and radiation designed into TVs, computers, and cellphones. Many states, municipalities, national governments, and regional blocs have passed laws that require safe disposal of e-waste as do a growing number of workplaces, schools, residential buildings, and neighbourhoods where green is slowly becoming the new normal.

All of us, through our teaching, activism and research, can contribute in some way to these efforts to press for a culture of sustainability over the prevailing one of consumerism—to advocate for a way of thinking and acting that is based on the idea that the Earth has limited resources to support human activities and limited capabilities to absorb and recycle our excesses. A culture of sustainability is also built on an ethics of intergenerational care with the enduring solidarity that binds our high-tech destinies to those of workers and the planet.

We should celebrate the organizations that have inspired so many green citizens to become involved and stay engaged in greening life. There's a growing list: NGOs like GoodElectronics and the Silicon Valley Toxics

Coalition work to keep digital devices from poisoning ecosystems in their place of manufacture. The Basel Action Network and the StEP-initiative relentlessly push for more extensive and thorough forms of end-of-life management of e-waste. Companies like Fairphone are pressing for ecologically sound design for high-tech goods and manufacturing processes that protect biophysical rights of workers who today fall sick from exposure to poisons and deadly factory operations.

It's time that these organizations did a better job reporting on their internal sustainability practices—where their procurement fits into the global supply chain; what their energy policies are; how they manage e-waste; and how they create model working conditions built on safe and sustainable uses of digital technologies. Give us another cause to celebrate being green.

## Notes

1. The nine boundaries are: 1. Climate change; 2. Change in biosphere integrity (biodiversity loss and species extinction); 3. Stratospheric ozone depletion; 4. Ocean acidification; 5. Biogeochemical flows (phosphorus and nitrogen cycles); 6. Land-system change (for example deforestation); 7. Freshwater use; 8. Atmospheric aerosol loading (microscopic particles in the atmosphere that affect climate and living organisms); 9. Introduction of novel entities (e.g. organic pollutants, radioactive materials, nanomaterials, and micro-plastics).
2. See for example, Basel Action Network, "Photos." http://www.ban.org/photos/
3. Stand with Standing Rock. http://standwithstandingrock.net/category/news/
4. Artists and Climate Change. https://artistsandclimatechange.com/about/
5. European Commission (nd). "Supporting Cultural and Creative Industries." http://ec.europa.eu/culture/policy/cultural-creative-industries/index_en.htm. Accessed Jan. 19, 2017.
6. Green Party USA (nd.). "Platform of the Greens/Green Party USA." https://www.greenparty.org/Platform.php. Accessed Jan. 19, 2016. Not even California's green party has acknowledged the material ecological problems of making culture, and this with evidence of massive pollution caused by LA-based film and TV production on its doorstep: Green Party of California (nd.). "Arts and Culture." http://www.cagreens.org/platform/arts-and-culture. Accessed Jan. 19, 2017.

7. Greenpeace (nd.). "Green IT at Greenpeace." http://www.greenpeace. org/international/en/campaigns/climate-change/cool-it/Green-IT-at-Greenpeace/. Accessed Jan. 19, 2017.

8. GoodElectronics. http://goodelectronics.org/. Accessed Jan. 19, 2017.

9. See work by Mark Graham and his colleagues on digital labor in Asia and Africa, for example, at the Oxford Internet Institute. https://www.oii.ox. ac.uk/research/. Accessed Jan. 18, 2017.

10. Human brains and neurological systems are intimately related to naturally occurring electromagnetism. The billions of neurons in our brains use electricity to function and communicate, as do all living cells, DNA, genes, and the rest of the building blocks of life. Birds and bees navigate by using the cryptochrome protein in their cells that sense the Earth's natural electromagnetic field. Disruption of this ability is said to factor in many disorders affecting these species. In humans, cryptochromes help set biological clocks. They tell our bodies when it's dark and when to sleep, two important triggers for the pineal gland's production of the hormone melatonin, which has important antioxidant properties and forms part of our immunological system, potentially helping our bodies fight cancer. Research suggests that cell towers and phones, and all electronics for that matter, are disrupting this natural biophysical rhythm. The Earth generates and maintains naturally occurring electromagnetism at frequencies that are relatively stable within a range of 3–60 Hz, with peaks of intensity at around 8, 14 and on up to 43.2 Hz. In our most relaxed conscious state, our brain waves operate in a frequency range of 8–12 Hz—the so-called Alpha waves—roughly the same fundamental intensity as the Earth's electromagnetic field. Our brains have evolved to resonate in very basic ways with the planet. Theoretical and Computational Biophysics Group, University of Illinois, Urbana-Champaign (nd.). "Cryptochrome and Magnetic Sensing." http://www.ks.uiuc.edu/Research/cryptochrome/. Accessed Jan. 19, 2017; Mohammad Ali Eghbal, Aziz Eftekhari, Elham Ahmadian, Yadollah Azarmi, and Alireza Parvizpur (2016) "A Review of Biological and Pharmacological Actions of Melatonin: Oxidant and Prooxidant Properties." J Pharma Reports 1: 106–115 (Eghbal et al. 2016).

11. In one common industry test, mobile phones are placed against a dummy's head, or Specific Anthropomorphic Mannequin (SAM), to measure where a phone can be safely held (you can find guidelines under the legal section in your phone's general settings). The SAM skull is modeled on large male heads, which tend to be bigger than average heads and, more importantly, thicker than those of vulnerable children and teen cell users. Consumers For Safe Phones (2011, Nov. 27). "FCC's Cell Phone Testing Dummy is Larger Than 97% of All Cell Phone Users." http://consumers4safephones.com/

fccs-cell-phone-testing-dummy-is-larger-than-97-of-all-cell-phone-users/. Accessed Jan. 19, 2017.

12. The question of "independence" of radiation research is vital. The Italian Supreme Court was the first to challenge the mobile phone industry's war-gaming tactic, when in October 2012 it ordered Italian authorities to pay workman's compensation to a former businessman who developed a tumor in his head because of heavy, long-term use of his mobile phone. The court threw out all the industry-funded studies on the grounds that they were tainted by conflict of interest. It accepted only the results of *independent research* showing a variety of causal links between mobile phone use and certain types of cancer. "Italy court ruling links mobile phone use to tumor" (2012, Oct 19). *Reuters.* http://www.reuters. com/article/2012/10/19/us-italy-phones-idUSBRE89I0V320121019. Accessed Jan 19, 2017; "ICEMS Position Paper on the Cerebral Tumor Court Case" (2012, Nov. 4). http://icems.eu/docs/ICEMS_Position_ paper.pdf?f=/c/a/2009/12/15/MNHJ1B49KH.DTL.

# CHAPTER 5

# Digital Desires: Mediated Consumerism and Climate Crisis

*Justin Lewis*

## INTRODUCTION

When people are asked to list the industrial sectors that are the most culpable contributors to climate change, most people would choose sectors like the oil, automotive or aviation industries. Some, aware of the problems associated with meat production, might opt for agriculture. But few people would nominate the media and communications sector, which is seen as light on its carbon feet, leaving few traces in its virtual wake.

Much of the research on the climate impact of information and communication technologies (ICTs) takes the same view. So, for example, institutions like the European Union and the World Bank point to the potential role of ICTs in *reducing* carbon emissions. A paper for the World Bank made a series of optimistic projections:

> The transformational potential of new information and communication technologies (ICTs) was on display in Paris at the Twenty-First Conference of the Parties (COP21) to the United Nations Framework Convention on Climate Change. ICTs—including the Internet, mobile phones, geographic information systems (GIS), satellite imaging, remote sensing, and data

J. Lewis (✉)
Cardiff University, Cardiff, UK
e-mail: LewisJ2@cardiff.ac.uk

B. Brevini and G. Murdock (eds.), *Carbon Capitalism and Communication*,
Palgrave Studies in Media and Environmental Communication,
DOI 10.1007/978-3-319-57876-7_5

analytics—could reduce yearly global emissions of carbon dioxide ($CO_2$) 20% by 2030 (Gallegos and Narimatsu 2015).

The ephemeral nature of many aspects of media and communications—with words and images passing unseen through cables and across the ether—makes these possibilities easy to imagine. Content—information, culture and entertainment—need no longer be contained in manufactured objects (on paper, in CDs or DVDs, for example) but can distributed digitally and directly to a wide array of electronic devices. It is easy to see how the World Bank and others might be enthused by this virtual potential.

But there is a problem. While digital technology offers exciting opportunities for carbon reduction, the media and communications industries are travelling down a very different road. They are not entirely to blame for this—although they are not entirely innocent either. They find themselves caught up in and integral to a set of consumerist practices that threaten to thwart any chance we may have of tackling climate change. These practices involve consumerist business models (notably planned obsolescence) and the promotion of a consumerist credo. They define progress in ways that rely on increasing rather than stabilising or decreasing global levels of production and consumption. In short, they operate to make catastrophic levels of climate change more likely.

## Consumerist Business Models

For a media and communications industry driven by commercial imperatives, the development of virtual digital distribution systems—such as the internet—has significant drawbacks. This is a sector in which the selling of content is made more profitable when that content can be manufactured into objects—whether books, newspapers, photographs, magazines, records, CDs or DVDs. The growth of digital online distribution upset this lucrative system of exchange, making it possible to download news, music and most forms of information and entertainment either cheaply (with rigorous intellectual property policing and firewalls) or for free.

This has obvious environmental advantages, although it is worth remembering that online digital communication is powered in ways that are not so different from traditional heavy industries. As Richard Maxwell and Toby Miller point out, digital objects and networks still have a physical presence that requires storage and capacity, which, in turn, needs electricity to power industrial plant in the form of servers (see Maxwell and Miller 2012). So, for example, a 2013 report from the Center for Energy-Efficient Telecommunications (CEET) described how the huge growth in wireless connectivity to cloud services was driving an increase in carbon footprint from six megatonnes of $CO_2$ in 2012 to up to 30 megatonnes of $CO_2$ in 2015, the equivalent of adding 4.9 million cars to the roads (Vertatique 2014). For most of us this storage is unseen, creating the misleading impression of a digital world that it is ephemeral and devoid of physicality.

Nonetheless, the reduction in the production and disposal of objects like books, newspapers, photographs, magazines, records, CDs or DVDs *should* enable a significant reduction in carbon emissions. The roadblock here is economic rather than technical: in a commercial world, the growth of virtual content is also bad for business. In media sectors like music and news, for example, it is difficult to monetise the virtual world, forcing those sectors to create new business models that will never be as profitable as those they replaced. The newspaper industry is hit with a double whammy: only a few newspapers have wealthy enough niche markets to successfully impose firewalls and charge people for online content, while advertising revenue for online editions is significantly less than it is for print. So despite increases in the amount of copy produced by journalists (Lewis et al. 2008), the industry has seen significant reductions in the number of newsroom employees over the last 10 years (McChesney and Nichols 2010). This continues apace—according to the Pew Research Centre (2014), who recorded a 10% drop in just 12 months (http://www. journalism.org/2016/06/15/newspapers-fact-sheet/).

The reduction in revenues in the music industry has been even more dramatic. In the 1980s, the industry's control over production and distribution allowed it to engineer a highly profitable shift from vinyl to CDs. The rise of the internet to become the main form of music distribution meant a significant loss of control. The shift from CDs and vinyl to downloads and streaming services, despite aggressive policing of intellectual property law, have had a profound impact on its profitability. While the consumption of music remains high, global revenues over the last 20 years have almost halved (http://ifpi.org/news/IFPI-GLOBAL-MUSIC-REPORT-2016).

For the non-commercial creative sector—the BBC, for example—this is not an issue, since their income does not depend upon selling content. If it can produce high quality, popular programmes while embracing principles of public service, it receives public funding. But for the commercial sector as a whole, their survival in a capitalist economy depended upon finding other ways to organise the routine production and consumption of objects. This has led to two equally damaging responses for the environment and our ability to tackle climate change.

The first—which I shall discuss in more detail in the next section—is an increasing dependence on advertising as a source of income. The second is a drive towards the *proliferation of media and communication devices* for the reception, communication, storing and playing of content. This proliferation is twofold: it involves an expansion in the range of devices that receive, send and play content, and a reduction on the lifespan of those devices. The first of these forms of proliferation has, at least, the merit of convenience: so, for example, while we may use a phone, a tablet, a laptop and a PC to do many of the same things (as well as more traditional devices like televisions, radios etc.), they each have their own benefits depending on where we are.

The second form of proliferation is more environmentally damaging and is generally *dis*advantageous for most consumers. Faced with a decline in sales of many cultural objects, the media and communications industry have, in a metaphorical sense, begun to turn hardware into software. Planned obsolescence is not a new idea in industrial production, not least in a media and communications industry that has made it a fine art (Sterne 2007). But it is now pursued with an unprecedented zeal, making the lifespan of communication and playback technologies increasingly fleeting.

The strategy for turning hardware into software is both physical and psychological. Obsolescence is built-into most electronic devices we now use. They are, in essence, a digital smorgasbords of components, some of which require renewing (like rechargeable batteries), some which may be regularly or occasionally upgraded (such as processors), and some relatively stable components (such as sound and image reproduction). But they increasingly come in one sealed container, so that, constrained by the inflexibility of its packaging, it will last only as long as its least durable component.

The mobile or cell phone—perhaps the most ubiquitous object of the digital age—is a conspicuous example of this kind of built-in obsolescence. Apple, a market leader in the higher end of this market, designed obsolescence into its devices at an early stage, making them difficult (using the notorious customised 'pentalob' screw) or impossible to open (Lewis 2016). Nearly all leading manufacturers have followed suit, replacing earlier models, which enabled you to change components, with sealed phones with a guaranteed limited lifespan. When Samsung's new phone was released in 2015 it was, unlike its forerunners, hermetically sealed. This 'upgrade' brought them in line with their competitors, meaning users will no longer be able to change the battery or upgrade storage. Consumers will now struggle to buy a phone that will last longer than its battery life.

This is now the industry standard. At the time of writing, the only phone company on the market that allows you to change batteries, faulty or outdated components is Fairphone, a social enterprise company based in Amsterdam, whose aim is to "create a positive social and environmental impact from the beginning to the end of a phone's life cycle" (https://www.fairphone.com/en/our-goals/). While a Fairphone will last considerably longer than any other phone on the market, it is not available at most retailers. Fairphone's survival is, in this sense, precarious, not only because of its marginality but because it has abandoned what is undoubtedly a profitable business model. Here we see a direct contradiction between *economic* sustainability in a consumerist economy and *environmental* sustainability.

The design of a phone used to be emblematic of a whole decade—today the average life of a phone is 12–18 months. The same temporality applies to most of the devices we buy—the cycle of replacement for TVs, laptops, iPads, gaming consoles or any other digital device has become bewilderingly brief. The idea that something should be built to last is, it seems, *so* last century.

The digital revolution has, in that sense, speeded up the production cycle. It has brought with it significant improvements—notably in computer power and capacity—and many of us can remember upgrades that were genuinely better than what came before. But these improvements are usually accompanied by mediocrity and flimsiness—the quality of sound on most gadgets, for example, is no better than it was in the Sony Walkman's of a bygone age. As the gurus of planned obsolescence know all too well, perfection can only be sold once.

The media and communications industry has also used marketing strategies to institutionalise obsolescence, to create the idea that in the digital age progress *depends upon* the regular replacement of devices. Again,

Apple has shown a particular flair for, in effect, celebrating the short life span of its products, turning each new product launch into a media event.

While ideas of human progress have, for some time, been linked to technological development, the fanfare that now marks the impending death of fairly new devices amounts to a small but significant ideological shift. We are asked to measure progress entirely in consumerist terms: rather than evaluate the extent to which a new technology advances the human condition—either individually or collectively—we are encouraged to see progress as a conveyer belt of product upgrades (Lewis 2013).

The damaging environmental consequences of this hyper-production cycle are its dirty little secret. We are used to measuring energy use purely in terms of individual consumption, aware of the resources it takes to power devices, but oblivious to the resources used in their manufacture and disposal. This conceals the real environmental costs of our upgrade economy. A report by the Restart project (2015) on the iPhone 6 indicated that:

> Eighty five percent of the estimated 95 kg of $CO_2$ emitted during its entire lifecycle (manufacture, use and ideally recycling) occurs at the manufacturing stage. The estimated manufacturing footprint of 80 million iPhone 6 projected to be sold (6,460 kilotonnes) will be greater than the total annual carbon footprint of the London boroughs of Westminster, Lambeth and Camden – of over 770,000 people and all of the business activity in three central areas of one of the world's richest cities. The iPhone 6 use-phase footprint, the energy used to charge and operate the mobiles—accounting for only 11% of those 95 kg—is almost insignificant in comparison.

Behind these slim and sleek devices are a series of heavy industrial processes in mining and manufacturing. So, for example, the average PC and monitor requires the same volume of resources to build as an SUV (Kuehr and Williams 2003) and the toxicity of the production processes mean that some of the most hazardous waste sites in the US lie beneath the fashionable surface of Silicon Valley (Gabrys 2007). Once built, new gadgets blaze brightly but briefly before being dumped on rapidly growing mountains of toxic e-waste.

The problem, in other words, has less to do with the power hungry nature of media and communications devices than with the business model that requires their rapid replacement. The World Bank's optimism about the role of ICTs in reducing carbon emissions comes crashing up against

the profligacy of this business model. It is also indicative of a larger contradiction—one many technologists tend to ignore—between an economic
system that requires endless and permanent consumer-driven economic
growth and the need to reduce our carbon output (Lewis 2013).

If media and communications technology is to have a role in tackling
climate change, it will have to be developed outside the market mechanisms that dominate most production in the twenty-first century. Our
commercial system has produced a legion of what Vance Packard (1960)
once called 'waste makers'. Only if *freed from commercial imperatives* can
this technology allow us to enjoy its digital benefits in ways that stress
sustainability and minimise wasteful production.

For social enterprises like Fairphone to prosper, governments need to
change the terms of trade. So, for example, legislators in France have
attempted to curtail the rampaging built-in obsolescence that characterises
commercial media and communications. The law requires manufacturers
to tell customers what the expected lifespan of products might be, and to
inform them how long spare parts would be available to repair them. In
2016, this will include a mandatory 2 year warranty, obliging manufacturers to repair or replace faulty appliances for 2 years after purchase (Burtt
2015).

This kind of intervention is an important first step. But it comes at a
price: in the long-term, the production of enduring technologies will never
be as profitable as the bounty provided by planned obsolescence. It also
means challenging the idea at the heart of consumer capitalism: the flawed
notion that prosperity—and happiness, fulfilment, success and security—is
measurable purely in terms of the accumulation of consumer goods. It is,
after all, this idea that enables such a widespread acceptance of planned
obsolescence. This challenge will not be easy: it runs up against the largest
and most pervasive creative sector on the planet—the advertising industry.

## CLUTTERING THE CLIMATE: ADVERTISING AS A WAY OF LIFE

In September, 2016, the advertisements that normally adorn the interior of
Clapham Common Tube station in South London were replaced by playful
pictures of pussycats. For the busy commuter, the change was only a subtle
shift in the urban visual landscape. Images of cats—those most beloved of
internet creatures—might be used to sell any number of things. While the
imagery was rather more coordinated than the hotchpotch of advertising

that generally clutters tube stations and other urban settings, it was well within the imaginative zeitgeist of a creative ad campaign.

But a closer look revealed two curious features. Most of the pictures contained no text and no logos—just cats peering out from a variety of angles across a blank white space, or the word CATS plastered across the ticket barriers. Those that did contain words simply revealed the acronym, the 'Citizens Advertising Take-over Service', with no hint of who they were or what they might be selling.

Could this be a clever campaign by an inventive ad agency, perhaps a deliberately enigmatic precursor to a more revealing product pitch? Since advertising has become the common vernacular of visual communication (Ewen 2001), both these possibilities seem plausible. But behind the coy display in Clapham lies a profound question about creativity in the twenty-first century. Imagine how our cultural environment might look if advertising creatives were set free from the constraints of consumer culture? Suppose the principal imagineers of modern life (Lewis 2013) were no longer obliged to weave their tales around sales, but could tell any story or paint any picture?

The Citizens Advertising Takeover Service (CATS) asks us to take just such an imaginative leap—to think what life would be like if advertisers could do more than just celebrate consumption. The idea was the brain-child of James Turner, a former TV producer who had worked for Greenpeace, to provoke thoughts about "a world where people value their friendships more than the things they own" and to ask "if there's a model for creativity that doesn't have brands as its patron"?

During his years at Greenpeace, Turner became aware that most creatives working in advertising agencies were *not* there because of any particular commitment to consumerism. They apply their creativity to the selling of goods because we have created a world where so much of our creative talent is in public relations and marketing. In order to capture this sense of possibility, Turner set up Glimpse—a collective of creatives drawn from advertising and graphic design dedicated to "making positive social change feel attractive to millions more people" (http://weglimpse.co/).

Cats in Clapham Common was the latest—and perhaps the gentlest—in a series of anti-consumerist takeovers of advertising space. Adbusters have an impressive back catalogue of beautifully crafted 'advertisements' designed to subvert the original messages and create (borrowing from Raymond Williams famous definition of culture) a "new structure of feeling" (http://www.adbusters.org/). In 2014 in New York, an augmented reality app called No Ad enabled people to aim their phone at an advertisement in the New York City subway system and see it replaced by a work of art. For ten days in 2015, the city of Tehran replaced all its billboards with artworks: an idea was not entirely devoid of promotional intent, since the campaign was aimed to encourage people to visit the city's art museums, but it was a temporary respite from a more overtly consumerist aesthetic.

Most rebellions against advertising have been more self-consciously subversive. Feminist graffiti in the 1970s used wit and black spray paint to call time on the rampant sexism that characterised so much advertising of the period. A more recent addition to this artistic/satirical tradition took place during the COP21 climate change talks in Paris. Many of the talk's sponsors were companies like Mobil and Volkswagen, who despite their promotion of fossil fuel production used the talks to 'greenwash' their image, to appear to be part of the solution rather than part of the problem. A group with the mischievous title 'Brandalism' replaced posters for some of COP21 sponsors, in order to "highlight the links between advertising, consumerism, fossil fuel dependency and climate change" (http://www.brandalism.org.uk/brandalism-cop21).

Brandalism's work may be more challenging than the arch cuteness of the Clapham Common takeover—both legally (the CATS campaign was crowd source funded, Brandalism's reclamation of public space is more covert) and politically (Brandalism names and shames, while CATS offers a softer critique). Both, however, recruit talent from the world of advertising to protest against a kind of corporate monoculture.

They contest the messages in advertising because advertising is everywhere. We now have commercials in our schools and on our clothes. They clog up—with increasing speed—nearly every form of communication we devise. Our dominant TV genre—in terms of sheer volume—is not comedy, drama or sport, but advertising. The average British viewer is now exposed to 48 TV commercials a day. Recent studies showed that around 40% of US TV time is now taken up by commercials (Lewis 2013).

Media that were once largely commercial free—from movies to the internet—are now full of commercial messages. There was a time when the only advertising linked to Hollywood films was the local cinema's pitch for the restaurant round the corner. All Hollywood movies now come with a slew of product placement deals or products tie-ins—which, in some cases, are more profitable than the box office take (Wasko et al 1993; Kretchmer 2004).

Not so long ago, musicians were reluctant to see their work endorse shampoo or sportswear. Today, the music and advertising industries are locked in a lucrative embrace (Klein 2009), with once rebellious punk rockers like Johnny Rotten and Iggy Pop fronting up TV commercials for butter and insurance. This may not be edifying but it is a profitable symbiosis: the ad promotes the music and the music promotes the product.

At the centre of this drift towards a consumerist monoculture is the internet. The beauty of the internet—a place where production is cheap and information is both abundant and free—has been shaped and constrained by a consumerist economy (Curran et al. 2012). Earlier media revolutions—notably broadcasting—were more contested, with many governments creating funding mechanisms for public service broadcasting that allowed content to be shown without commercials. Although some public service broadcasters have a significant online presence (the BBC website, for example, is by far the most the popular source of online news in the UK) most content online is a form of public relations or paid for directly by advertisers.

Our main search engines (Google) and social media (Facebook) are not only replete with commercial messages, they are part of increasingly sophisticated developments in advertising strategy, enabling advertisers to target consumers with hitherto undreamed of precision. Google searches, for example, are driven as much by market logic (who pays, where the consumer traffic is) as by closeness to content, while Facebook encourages subscribers to provide a slew of demographic details which can then be sold on. Even critiques of advertising posted on You Tube (including my own TED lecture) now come preceded—and surrounded—by commercials.

Advertising thereby sucks up a huge portion of the world's talent for art, design, creativity and storytelling. It has become such a routine part of everyday life that—unless prompted by CATS in Clapham or Brandalism-style interventions—we rarely stop to think about the cultural and political assumptions contained in most advertisements, or indeed, their ubiquity in our cultural environment.

This matters, because for all their diversity, advertisements share one basic value system. While Toyota, McDonald's and Tesco all have their own pitches, they sell more than just goods. Collectively, they sell consumerism as a way of life. The common thread that runs through all product advertising is that the only way to happiness, satisfaction, security or popularity is through consumption. The answer to every problem lies not in our relationship with friends, family or community, but in the dead world of things. Hole in the ozone layer? Buy factor 50 sun-cream. Not popular enough? Buy a new phone. Feel you need more control over your life? Buy a new car. Worried about your health? Consume this food or these drugs.

Most advertisers know that bound up in these carefully contrived associations are a series of fictions. In wealthy countries people's quality of life is closely linked to social relationships, health, security and community rather than their involvement in consumer culture (Jackson 2010; Kasser 2002; Wilkinson and Pickett 2009). For this reason, most ads spend little time extolling the use value of the object being sold, attempting, instead, to create a mythic association (in Roland Barthes' sense of the word) between the object and the social world (Jhally 2006; Lewis 2013).

These messages are capitalism's balm, countering concerns about pollution, climate change and exploitation. Any fears can simply be greenwashed away. Most ads keep us so focused on the pleasures of consumption that the people who actually *make* the endless array of commodities become irrelevant. Campaigners for fair trade have had to battle against this myopia, to force people to look beyond the beguiling images of buying. But it is a constant struggle: for every time our attention is drawn to sweatshops and the low wages that keep prices down and profits up, there are a thousand soothing messages that whisper, 'just buy it'.

If we are to tackle climate change, this is, to say the least, an unhelpful cultural environment. Climate change requires us to think collectively, advertising encourages us to act as individuals. Climate change means investing in public, low carbon infrastructure, an act of will and imagination that is simply outside advertising's field of vision. Climate change means understanding the environmental costs of the production and disposal of all the things we consume, advertising focuses purely on the moment of consumption.

Perhaps most conspicuously, the task of reducing carbon emissions will be difficult even if we keep consumption in the rich developed world at current levels. Advertising—relentlessly, hundreds of times every day—encourages us to consume more. Tim Jackson (2010) points out that with even small increases in consumption and growth, emissions targets look ever more beyond our reach.

Indeed, since it is difficult to begrudge increases in consumption and economic growth in poorer, developing countries, the responsibility to avoid hyper-consumption in the wealthier parts of the world (where it will have little impact on people's quality of life; Wilkinson and Pickett 2009) becomes ever more pressing. And yet, advertising focuses its resources on those people who already have the most, since that is where most of the world's disposable income resides.

Advertising is, for all these reasons, a deeply political discourse, one that is antithetical to tackling climate change. It confuses any attempt to question the benefits or to count the environmental costs of consumerism. In the developed world, curtailing consumption may have little impact on most people's quality of life, but it is an extremely hard sell. And it is the constant presence of advertising in every urban and media space that makes it so.

## CREATING A NEW CLIMATE

The picture painted thus far is a pessimistic one. The media and communication industries are a dominant part of most people's lives—in countries like the UK and the US, the average person spends over 8 hours of every day using, watching or listening to a media or communication device (Lewis 2013). When the production and consumption of those devices is dominated by a business model that encourages profligacy and waste, and when their content is dominated by a discourse promoting consumerism, any optimism about their role in reducing emissions seems misplaced.

This has nothing to do with the technology, whose potential remains exciting and socially beneficial. The problem is an economic one—principally a market system that relies on profligacy for profit and where the easiest way to fund content is through advertising. And yet the system manifestly fails on its own terms.

Planned obsolescence is bad even for the technologically sophisticated consumer, forcing them to pay for components they don't need. Advertising may be the dominant genre across most media, but it is the only one that exists regardless of consumer demand. In the UK, when the same event is broadcast on both non-commercial (BBC) and commercial channels, most people choose to watch without commercials. Almost by definition, advertising repeatedly fails on its promises and creates dissatisfaction (Kasser 2002): if buying certain products really did make us happy, fulfilled, popular, healthy and secure, then that would be the end of it. But there is no end point in advertising, which keeps on coming, in wave after wave, so that we are always on the edge of fulfilment without ever being able to reach it.

Meanwhile, in the more potentially civic media spaces—notably the news media—climate change has become old news, with coverage reaching the levels of a minor news story that is neither urgent nor pressing (Daly et al. 2015) This is, in part, because of the news media's own commitment to planned obsolescence (Lewis 2016).

In the twenty-first century developed world, these failures have become endemic. But they are also grounds for optimism. Both planned obsolescence and the superfluity of advertising are intrinsically unpopular. Breaking the link between consumerism and climate change will not be easy, but it does have the basis for a popular mandate. And social enterprises like Fairphone, Glimpse and Brandalism offer us glimpses of a different, more sustainable cultural economy—one that is a necessary condition for questioning consumerism and addressing climate change.

# CHAPTER 6

# From "Waste Village" to "Urban Circular Economic System": The Changing Landscape of Waste in Beijing

## Xin Tong

### INTRODUCTION

The growth of municipal solid waste (MSW) has become a pressing environmental challenge in many cities in China. As in other countries, the increasing demands for waste disposal, the rise of not-in-my-backyard (NIMBY) objections to facilities close to residential areas and the shortage of sites for landfills are prompting a search for alternative strategies to solve the problem (Davoudi 2000). Among these, the 3R concept (reduce, reuse and recycle) has been promoted in China as a key component in the national strategy for constructing a "Circular Economy" (Yuan et al. 2006). However, the evolution of the urban waste management system has proved to be a typically wicked problem for contemporary Chinese society.

Social approaches or technical solutions have proved to be inseparably bound up with a broad range of social, economic and environmental issues, prompting clashes of conflicting values among the various agents and decision makers involved (Churchman 1967). Moreover, both the definition of waste and social acceptable solutions for waste disposal have

X. Tong (✉)
Peking University, Beijing, China
e-mail: ongxin@urban.pku.edu.cn

© The Author(s) 2017
B. Brevini and G. Murdock (eds.), *Carbon Capitalism and Communication*,
Palgrave Studies in Media and Environmental Communication,
DOI 10.1007/978-3-319-57876-7_6

changed over time (Strasser 2000). This has involved more than providing technical solutions for the disposal of physical detritus. Increasingly critics have been targeting the mode of mass production and consumption and the consumer culture it supports (Featherstone 2007). Furthermore, there has been growing popular sympathy for the marginal populations working in this informal economy sector in miserable conditions (Medina 2010). These expanded definitions of the waste problem have led us to take a more inclusive view of current solutions and proposals for improvement, among which community-oriented tools have received increasing attention in both theory and practice (Zotos et al. 2009; Weinberg et al. 2000; WorldBank 1999).

In this paper, we offer tales of two villages with high concentrations of urban scavengers and recyclers—Bajia in the Haidian District, and Dongxiaokou in the Changping District—both located in the northern rural-urban fringe of Beijing. In the next section, we place the story of these two communities within the "confusing ramifications" (Churchman 1967) of the evolving urban waste management system as it has developed during the dramatic Chinese social-economic transition that started in the 1980s. In the section entitled "The Illusory "Urban Circular Economic System"", we analyse the community-oriented strategies used by the municipal administration in the design of the "urban circular economic system" since 2000, exploring its linkages with the waste villages and its consequences for landscape change. In conclusion, we call for constructive answers from the urban planning system to the challenge of waste by suggesting some principles of design for recycling in community public spaces.

## TALES OF TWO "WASTE VILLAGES" IN BEIJING

The economics of waste has changed drastically in China over the past decades, so have the practices of recycling and reuse in people's daily life (Goldstein 2006). The growth of waste in urban China is a by-product of fast industrialization and urbanization. The rapid transition of urban lifestyles towards mass production and consumption poses a major environmental challenge for Chinese contemporary cities (Yang et al. 2014; Xiang et al. 2011). As a result waste has become embedded into the collective life of modern cities, not only creating unpleasant landscapes, but also producing spaces that embody social and environmental injustice.

Waste villages are places of particular tensions and conflicts in urban development. They have received increasing public attentions since the late

1990s as marginal spaces dramatizing multiple problems associated with China's social transition (Tang and Feng 2000). Existing literature has addressed the economic linkages between the emergence of waste villages and the transition of the urban recycling sector from a planning system to market forms of organisation. Under central planning China used to have a comprehensive waste collection and recycling system in which recycling activity was presented and celebrated as central to the socialist economy's resource conservation actions (Goldstein 2006). This system collapsed during the transition from a planned to a market economy, along with the social values attached to it, starting from the beginning of the 1980s.

The consumption behaviour of urban citizens has changed dramatically in the last three decades, and the urban recycling sector has become increasingly integrated into junk-buyer networks mainly consisting of migrant rural workers. A spatial division of labour has emerged along the chains of recycling and waste disposal activities, from the urban pickers on the streets, to the junk sorting and local markets for recyclable goods, to the recycling clusters in rural areas decentred from urban centres, where recyclable materials are further processed into secondary materials or products (Li 2002; Linzner and Salhofer 2014).

The two waste villages analysed in this paper are concentration sites of junk sorting and markets on the rural-urban fringe, which is the pivot between rural and urban along the constant conversion chains of waste/recycling activities across the country. According to Liu and colleagues (Liu et al. 2008), there were more than 120 concentrated markets for recyclable goods in Beijing in 2004 with total land areas up to 240 ha. Among these only 24 had licenses from municipal authorities. The majority were on land rented from villages near the city, attracting large number of urban scavengers to do junk sorting nearby.

Our study focuses on two sites among the waste villages existing in Beijng: Bajia and Dongxiaokou. Field studies were conducted between 2007 and 2013. We witnessed the demolition of Bajia around the time of the Olympic Games, and the struggle against the demolition of Dongxiaokou, the biggest urban recycling centre in Beijing, in 2011. To investigate their linkages with the city, we conducted on-site interviews with more than 100 urban scavengers scattered in different residential areas in Beijing in 2012, observing their working conditions, and discussing possible public-private cooperation in community-based recycling projects. This work, which is part of a project for building an enhanced e-waste management system in China, provided us with opportunities to access different stakeholders in the system, including migrant rural recyclers and local administration agents, and to think about the changing landscape of waste through participant observation.

## *Spatial Shifts*

In the last decades, the location of waste villages in Beijing has been shifting outwards as the urban built-up area has been expanding. As the experience of Bajia and Dongxiaokou demonstrates, waste villages have become a special landscape that recurrently flourishes on the rural-urban fringe, which is then devoured by the expanding city, and moves outwards re-emerging in the shape of a new, even larger, waste village at the new frontier of the urban sprawl.

Bajia was a village in the Haidian district in the north of Beijing with an area of 1.6 square kilometres. In 1992, the former waste village in Erlizhuang, to the southeast of Bajia, was demolished due to the urban expansion. At that time, there was a recycling station in Bajia run by the local government of Dongsheng Town. As a result, many urban scavengers moved there, renting county yards from local peasants and starting junk sorting and recycling activities. By the middle of the 1990s, more than 5000 rural migrants were working in the recycling sector in Bajia, surpassing the number of local residents. Because more than 75% of these migrants came from Henan province (a populous province in central China with large outflows of rural workers), scholars, in the field of migrant labour studies, named Bajia the "Henan Village" (Tang and Feng 2000).

Because it is close to Zhongguancun, Beijing's Silicon Valley, Bajia has gradually specialized in e-waste recycling since the early 2000s. Migrant scavengers with e-waste recycling experience in Guangdong province were the first to develop this specialism. Later, more people became aware of the business opportunities and entered the market. Because several large IT markets, universities and R&D institutes are located in Zhongguancun, the amount of discarded computers and peripherals increased quickly. Recyclers used to collect e-waste from end users. Products that were usable or repairable were refurbished and then resold in the secondary market. For those unusable, recyclers used to disassembled them into plastic, glass, metal, wire and electronic components, and sell them to the recyclers outside Beijing for further processing. The e-waste recycling business was so profitable that, till 2007, there were more than 20,000 migrants flooding into Bajia.

On the one hand, the concentration of migrant recyclers was a good source of income for local peasants, who rented them their land. On the other hand, the high density of new residents combined with the recycling activities and storage of recyclable goods made the village crowded and chaotic. As the city expanded however, the surrounding areas were developed, land values increased, and demolition started in Bajia. As a result, in 2008, more than half of the migrant recyclers had to move out to find a new place for their business. Dongxiaokou became the most popular destination.

Before 2003, the Beijing Administration for Industry and Commerce was in charge of supervising and giving approvals to junk markets. Only entities that had received a certification could operate businesses. However, in 2003 this system was abolished and new junk markets appeared spontaneously. The first was established in Dongxiaokou in 2003. The former waste village in Wali, which was closer to the urban centre, was demolished and the Beijing Olympic Park built at its place. A local peasant in Wali, who had been doing junk sorting and recycling since the 1980s, decided to rent a 33 acre land plot in Dongxiaokou and established a junk market. He divided the market into several zones. The surface of each zone was 50–100 square meters and had a small cottage to live in and a yard for junk sorting and storage. He rented out the entire area to migrant scavengers, providing them with a relatively comfortable place to stay. Other markets established later followed this model.

As a result, Dongxiaokou attracted more than 700 family workshops specialized in different kinds of junk recycling, such as plastics, PET bottles, cotton fabrics, waste electronics and furniture. Every recycling zone was located according to its category of recyclable goods. A member of the local government said "Here, you could find everything used in your home, being sorted and recycled by specialized workshops". Gradually, Dongxiaokou became the largest waste village in Beijing, absorbing more than one-quarter of recyclable goods produced in the city.

However, the concentration of urban recyclers in both Bajia and Dongxiaokou increasingly attracted negative comment in public media. Objections came from local residents who complained about the dirty and congested environment. Environmentalists scaled-up the local environmental concerns to address issues of uncontrolled flows of recyclable goods to other regions and highlight inappropriate ways of processing, hazardous for the environment at large.

Although junk market operators rented land from local villagers, its use for waste markets and workshops was formally illegal. According to the Chinese Land Administration Law, rural land cannot be developed for activities other than agriculture without previous expropriation by the Municipal government and a change in its function to urban status. Again, in 2011, demolition came to Dongxiaokou. However, this time the scavengers attracted support from a local NGO, which raises public awareness of their marginal and unequal status pointing out that while they provided low cost recycling services to the city they could hardly get a place of their own for living. Partly due to their efforts, some of the junk markets were temporarily reserved and the local government promised to provide another place to accommodate those migrant workers.

## Social Restructuring

The spatial shifts of waste villages in the Beijing metropolitan area were accompanied by a social restructuring within the floating scavenger population, which Tang and Feng (2000) have described in detail. Their study focuses on the social relations and mobility of migrant scavengers in Bajia after their migration from their rural hometowns during the 1980s and 1990s. They highlight the way that the transition from a planned to a market economy created a disjunction between rural and urban societies, and an occupational stratification between urban and rural workers, which eventually resulted in discrimination against rural migrant workers without technical skills or business capital who were often relegated to the rank of 'underclass' citizens. Faced with this situation, they argue, migrant workers took to rag picking as a way of working in Beijing without social or economic resources. However, as Tang and Feng (2000) also point out, a significant stratification also appeared among scavengers rooted in a hangover from the administrative structure of the former central planned recycling system, which was in place before the market transition of the 1980s.

Established in 1965, Beijing had a three-level collection and recycling scheme. The whole system was under the control of only one state-owned enterprise. This used to manage sites and stations according to economies of scale. Recyclable products were collected from households and firms through different channels, thanks to an advanced multi-site network based in both streets and local communities. Collected waste was then shipped to recycling stations situated in every district and county and finally transferred to state-owned enterprises for reusing and recycling treatments.

During the market transition the central planning system was dismantled and state-owned enterprises, across the whole country, gradually abandoned junk sorting and recycling. This was in part due to the scarcity of financial resources but also to the lack of personnel, who were increasingly attracted by more profitable jobs. The migrant rural workers seized this opportunity and took the place of the community collection sites by peddling door to door, collecting discarded goods from households, or picking rags on the streets. Starting from the very bottom, some migrant scavengers accumulated enough capital to move up the recycling value chain. They rented rural land at the rural-urban fringe to buy recyclable goods from other peddlers or door-to-door scavengers. However, this kind of business was illegal until 2003.

Following the abolition of the administrative approval system for junk markets in 2003, which eventually ended the monopoly of state-owned recycling companies, the number of junk markets increased rapidly, from the original 23 state-owned recycling stations to more than 120 junk markets in 2004 (Liu et al. 2008; Shang 2006). Many of the state-owned recycling stations suddenly confronted increasing competition from private businesses run by migrant recyclers.

Dongxiaokou was one of the junk markets in Beijing, which grew at the fastest rate since 2003. This was due to its convenient location and its efficient organization for migrant recyclers. There were seven major markets run by different managers. Six of them came from Henan province. Furthermore, more than 90% of the residential recyclers in Dongxiaokou came from the same area in Henan. They depended on kinship networks to run the whole chain of junk recycling. This included collecting on the streets, sorting in small yards in Dongxiaokou and delivering final materials to their customers outside Beijing, such as paper mills, plastics plants, and metal workshops located in Hebei, Henan, Shandong and other cities. Some recyclers even owned their own plants for processing secondary materials and adding value outside Beijing.

Following the government decision to demolish Dongxiaokou in 2011, the majority of migrant recyclers accepted the new reality and looked for new places to move to. However, the market managers tried to contest the decision through different channels. One manager from Henan, who ran an iron and steel scrap market in Dongxiaokou, collaborated with a TV channel in Shanghai to produce a documentary video on waste village migrants' workers daily lives, on their struggle for living, and on their contribution to the city and to resource recovery (Real-25-hour 2012). The video also shows his effort to upgrade the recycling industry and improve working conditions in Dongxiaokou. However, his publicity effort did not stop the demolition of the market.

## The Illusory "Urban Circular Economic System"

Since 2000, along with the demolition of waste villages, the municipal government has been trying to rebuild the urban recycling system. This has the aim of replacing the informal scavengers' network with an extensive set of community-based collection facilities covering all residential areas. Several large-scale collection and sorting centres equipped with automatic machines were planned to replace the dominant labour-intensive sorting activities of the waste villages. However, the implementation of this plan has confronted constant difficulties (Wang and Han 2008).

### Community-Based Recycling: Social Approaches

In May 2000, nine administrative departments of the Beijing Municipal government jointly launched an agenda for a pilot project establishing community-based recycling systems. On the one hand, the project aimed at rebuilding the urban recycling system in the inner-city. Five of the eight inner-city districts of Beijing, Xicheng, Chaoyang, Haidian, Fengtai and Xuanwu, were nominated as pilot regions. The actions included promoting the standardization of the logos, transportation vehicles, workers uniforms, prices and categories of recyclable goods, as well as measuring equipment in the recycling sector. On the other hand, the government planned an extensive network of community collection sites, each serving 1000–1500 households. More than 1800 sites were designated covering all the districts of Beijing inner-city. The government also planned to build 10 formal recycling markets before 2003. This scheme was supposed to cover 100% of recyclable products in the five pilot districts, to regulate the uncontrolled flows of rags and to reduce concentration of migrant workers in waste villages.

This plan can be considered as an effort to rebuild a three-level recycling system similar to the former centrally planned one. However, the emphasis went far beyond resource conservation, and pursued the goal of producing a neat and clean image for the urban Recycling sector. It imposed detailed requirements on community-based recycling facilities, including the design of cabins, logos and the appearance of the vehicles. It also addressed the further processing of recyclable goods, including sorting, compressing, cleaning and purification. A closed-loop urban recycling system was designed to fit the image of a modern international metropolis.

The plan also tried to address social problems in the city, including unemployment among disadvantaged groups. It assigned local unemployed workers priority as operators of community recycling facilities. Only if the company could not hire enough workers from among local residents were they permitted to hire qualified migrant labourers instead. The plan created a set of rules and standards for training, registration, social security, and human affairs management. It even suggested building a national qualification certificate system for workers in the sector.

The recycling companies inherited from the former state-owned recycling stations carried out the plan. In 2008, 20 recycling companies of this kind established more than 3000 community recycling facilities covering over 70% of the local resident population. They also built 13 sorting centres located in different districts. However, collection still relied on migrant scavengers. Although they provided cabins and uniforms to contracted workers, the formal recycling companies could hardly create a monopoly over such activities, either for them or their contractual employees. As a consequence, contracted recyclers still had to compete with informal scavengers working door-to-door. On the other hand, formally contracted recyclers preferred to sell their collected waste to the market that could offer the highest price, rather than at a fixed price to the recycling company they were contracted to. As a result, the sorting centres of the formal recycling companies were unable to acquire much of an advantage over their rivals in the waste villages.

### Reverse Logistic System: Efforts for Technical Optimization

Besides adopting a social approach in building community-based recycling models, the municipal government also tried to transform the recycling sector through optimization of the whole logistic system. In 2006, the Beijing Municipal Bureau of Commerce, together with 10 other municipal bureaus, announced a new version of the Pilot Plan for Promoting the Recycling Industry. This version put the emphasis on logistics and processing. The main aims were to reduce the number of street migrant scavengers and replace the recycling markets in waste villages with 10 big sorting centres equipped with automatic processing machines. In order to promote the upgrading of the urban recycling industry from the dirty and labour-intensive business of rag picking to a clean and efficient industry the municipal government established a special fund to support formal recycling companies in upgrading their equipment and enhancing their control over the whole recycling chain.

However, this optimized logistic system still found it difficult to compete effectively with the informal sector. Waste electronics is a prime example. The informal system, composed of peddlers in residential communities, mobile collectors and informal disassemblers in waste villages, such as Bajia and Dongxiaokou, at the rural-urban fringe, enjoy important advantages over the formal system for several reasons. First, the peddlers working in residential communities provide households with a door-to-door service, which is more convenient to customers. Second, their transportation tools are flexible and able to fit different conditions. Because peddlers in residential communities use tricycles for household collection they can reach out-of-the-way places in the urban sprawl of high density housing without incurring fuel costs. There are many mobile collectors who ride around communities and buy products from residential community peddlers. This decentralised, multi-level system, moves discarded products from dispersed households to markets very efficiently. Thirdly, the processing of used goods is also more flexible. Some discarded products are refurbished and sold to rural residents and poor migrants, and those that cannot be reused are sorted and disassembled during transportation. Fourthly, the materials are sold outside Beijing for further processing. Finally, the automatic capital-intensive sorting equipment is inefficient compared with the labour-intensive sorting of waste villages. Additionally, through the whole process, the informal sector pays no transaction taxes giving them a further cost advantage.

## Waste as Wicked

The effort to rebuild the recycling system to replace the informal sector has therefore by and large been in vain. On the one hand, the formal system is increasingly evolving towards a capital intensive waste management system with expensive technical solutions and even though the government is willing to pay the cost of implementation, by founding facilities for example, these projects face a growing number of NIMBY movements protesting the arrival of waste management facilities in their 'backyards'.

On the other hand, the cheap recycling service provided by migrant recyclers creates severe environmental and social externalities, such as environmental degradation and health hazards in waste villages, which continue to expand despite opposition from the local administration.

Waste villages illustrate the persistent problems of China's transition. The urban administration mainly approached the waste problem from the perspective of environment and resource conservation and failed to take full account of the social situation of the people who worked in this sector, but who could not be accepted as local citizens, and the tensions this exclusion generated. Nor did they confront the push to acquire land for profitable development that often underpinned the demolition of waste villages, concealed behind complaints over unpleasant social-environmental conditions.

As the urban centre has expanded rapidly the problem of waste villages has become increasingly firmly embedded within the deep urban-rural segmentation of the Chinese society, which structures the complex social dynamics of contemporary transition. As a result we would argue, the waste problem will remain highly resistant to solutions if the actors involved only approach it with a limited technical perspective of optimization.

## DISCUSSIONS: ADAPTIVE OR INCLUSIVE?

The paradox of wicked problems is that strategies for solving them are informed by the way one looks at them (Rittel and Webber 1973). The construction of an "urban circular economic system", which focuses on capital intensive recycling facilities and logistic infrastructures is based on "linear approaches and partial skill sets" (Xiang 2013). Effective action needs to take account of the complex realities of diverse waste management processes, and multiple forms of reuse and recycling.

As wicked problems have no 'stopping point', we believe that 'additional efforts might increase the chances of finding a better solution' (Rittel and Webber 1973). Community-based tools call for urban designers to review recycling activities as a part of a complex web of social interactions in the common space of the city. In our research, we witnessed positive interactions between migrant recyclers and residents in some communities, while tensions existed in others. Those recyclers who secure an independent space of their own in the residential communities are more confident and relaxed when interacting with other people. In designing such spaces, we need to stop seeing recyclers as 'the other', alien and separate from the local community, to whom we have to 'adapt', and consider them instead as a part of us, and create an 'inclusive' place for interactions, which will transform the isolated waste villages into cells of an urban circular economic system.

**Acknowledgements** This research was supported by grants from the China National Natural Science Foundation ('Technological Transition through Extended Producer Responsibility in Electronics Industry' [41271548]). We thank all of the interviewees for sharing their ideas with us during our field studies. Thanks to all anonymous referees for their advice and suggestions on the early draft.

# CHAPTER 7

# Big Data, Open Data and the Climate Risk Market

*Jo Bates*

As the deep structural "uncertainties" that are beginning to define the early twenty-first century continue to unfold (Hay and Payne 2013), the question of how societies should respond to the decline—and the consequences of—the era of carbon capitalism become increasingly pressing. For some, the answer to these challenges is found in a further defining trend of the contemporary era—advances in digital information and communication technologies such as big data analytics, smart cities and social media communications. In this chapter, we critically examine some key developments at one site where three phenomena related to these trends—climate change, big data and financial capitalism—intersect: data-driven climate risk markets. Situating these developments in the context of emergent forms of "informational power" (Braman 2006) and data policy struggles that aim to make meterological data more readily exploitable by climate market actors, the chapter asks what it might mean to turn to climatic uncertainty as a source of profit and growth.

J. Bates (✉)
University of Sheffield, Sheffield, UK
e-mail: jo.bates@sheffield.ac.uk

© The Author(s) 2017
B. Brevini and G. Murdock (eds.), *Carbon Capitalism and Communication*,
Palgrave Studies in Media and Environmental Communication,
DOI 10.1007/978-3-319-57876-7_7

## CLIMATIC UNCERTAINTY

As many scientists and commentators have observed, planetary ecosystems are in a state of crisis. In a 2009 *Nature* article, Johan Rockström and colleagues identified the various ways in which human action is stressing the ecological "carrying capacity" of the planet (Rockström et al. 2009a). In their paper, the authors quantify a range of "planetary boundaries" in order to identify a variety of ecological processes and the "associated thresholds" that could not be crossed without generating "unacceptable environmental change". Their analysis identifies that the boundaries for "climate change, rate of biodiversity loss and interference with the nitrogen cycle" have already been passed—unacceptable environmental change is occurring. Further, in the case of "global freshwater use, change in land use, ocean acidification and interference with the global phosphorous cycle", these boundaries are being quickly approached.

Such warnings about the impact of human action on the Earth's ecosystems are echoed by many expert commentators, including scientists involved in the Intergovernmental Panel on Climate Change (IPCC). The IPCC assessment reports are a collaborative effort involving thousands of researchers and governments from around the world. The report is a systematic review of publications relevant to the scientific, technical and socio-economic aspects of climate change. The aim is to provide a comprehensive view of current knowledge. The most recent report was published in 2014, and its conclusions were unambiguous:

> Warming of the climate system is unequivocal, and since the 1950s, many of the observed changes are unprecedented over decades to millennia. The atmosphere and ocean have warmed, the amounts of snow and ice have diminished, and sea level has risen.

Further,

> Anthropogenic greenhouse gas emissions have increased since the pre-industrial era, driven largely by economic and population growth, and are now higher than ever. This has led to atmospheric concentrations of carbon dioxide, methane and nitrous oxide that are unprecedented in at least the last 800,000 years. Their effects, together with those of other anthropogenic drivers, have been detected throughout the climate system and are extremely likely to have been the dominant cause of the observed warming since the mid-twentieth century (IPCC 2014).

## BIG DATA AND INFORMATIONAL POWER

Scientific knowledge about phenomena, such as climate change, is highly data dependent. Without the appropriate data it would not be possible to identify the extent to which the global climate has altered over the years or analyse the potential causes. Without the ability to analyse vast amounts of weather observation data, society would only experience the consequences of climate change: extreme weather, rising sea levels, crop failures and so on.

The meteorological and climate sciences have long been data-driven disciplines (Edwards 2013). The reason we know that the climate is changing, why it is changing, and how we should respond, is the result of decades of complex processes of data collection, cleaning, analysis and modelling by climate scientists. Meteorological organisations around the world hold vast archives of weather observations that can be analysed to predict the weather and understand average weather conditions—the climate—over time. Scientists are also exploring innovate ways to fill in the gaps in their datasets in order to increase their understanding. For example, citizen science projects such as Old Weather (https://www.oldweather.org/) that use the labour of volunteer 'citizen scientists' to transcribe historical shipping records, so that the digitised data can be fed into weather observation databases and climate models.

However, it is not only scientists that are interested in using data in order to understand and respond to changes in the climate. As meteorological data becomes more abundant and fine-grained, it—like many other forms of data—is being increasingly exploited by those who want to use high level data analytics in order to extract profits. This, according to Mayer-Schönberger and Cukier (2013), is the era of "datafication" in which more and more aspects of human existence are being quantified and turned into computerised data. Over the last decade it has become almost cliché to claim that "data is the new oil", or—as the former European Commissioner for Digital Agenda Neelie Kroes' (2011) claimed—"the new gold". Exponential increases in data and computing power, the World Economic Forum argue, are fuelling a "Fourth Industrial Revolution" (Schwab 2016). While such claims are open to critique, for example, we may draw on Webster's (2006) analysis of earlier claims about the revolutionary nature of informationalisation that question the notion of revolutions *within* a capitalist political economy, it is still vital to recognise the deepening dependence of the capitalist mode of production on data.

Sandra Braman (2006) alludes to these developments in her conceptualisation of the "Informational State". In her work, she observes the development of a deepening form of "informational power" beginning in the 1970s and 1980s. She argues that while analyses of power have tended to categorise the concept into instrumental, structural and symbolic forms of power, processes of information intensification in recent decades have brought a further type—"informational power"—to the core of contemporary power relations. This "informational" form of power, she argues, interacts with other forms of power by "manipulating" their "informational bases" (p. 26). She illustrates a number of examples of this developing "informational base" for instrumental, structural and symbolic forms of power with reference to Smart Weapons, internet surveillance, personalised web-services, social profiling and manipulation of public opinion. Braman further argues that the processing and distribution of information are also often key factors in "the transformation of power from potential to actual" (p. 27).

It is important that this informational form of power is addressed as we try and navigate through the complex and uncertain terrain of the contemporary era. It is clear that data analytics will contribute to how societies respond to the significant challenges we face in the twenty first century, but also that many of these data-related practices generate deep uncertainties of their own. The different interests that are empowered and disempowered by how data are generated, processed and used will be heavily influenced by the wider dynamics of political economy and culture. It is therefore important to integrate an understanding of datafication into our analysis of the broader processes of change. In order to 'make real' some of these issues, the rest of the chapter will now turn to examining such developments in relation to climate risk markets.

## The Climate Risk Market

Similar to climate science, financial markets have long been data-dependent. Streams of data are crunched by algorithms and models, and feed into human and automated decision-making processes which have significant impact throughout society. Weather derivatives are a type of climate risk product traded in the global financial markets. These financial products cover businesses for "moderate departures" from expected weather conditions, as opposed to traditional indemnity insurance, which covers for significant departures (e.g., extreme events) and catastrophic loss

(Dischel 2002). Rather than insuring against a specific observable loss, payouts on weather derivative contracts are instead triggered when particular meteorological conditions, e.g., the average temperature over a month, are detected in vast indexes of weather observation data. Key actors in the markets include businesses and governments wanting to hedge against climate risk, re-insurance firms, institutional investors, and exchanges such as the Chicago Mercantile Exchange (CME).

Much of the primary market trading in weather derivatives occurs in the over-the-counter (OTC) market, through which bespoke contracts are negotiated in private between buyers and sellers (Speedwell Weather n.d.). Buyers typically are firms in sectors such as energy, agriculture and construction; while sellers tend to be re-insurance firms such as SwissRe. There is also a secondary market in weather derivatives that trades primarily through the CME (SCOR Global 2012). In this secondary market, primary market contracts are traded in order to manage risk. While many buyers in the primary markets have traditionally been aiming to hedge against climate related risks, a new class of speculative investor in climate risk has emerged post-financial crash. In 2013, the largest source of new trades in the market was from hedge funds speculating on average monthly temperatures (Thind 2014), which essentially means using data-driven techniques in order to speculate, and ultimately profit, on climatic uncertainty.

Weather derivatives were developed within the US energy industry by Enron, Koch Industries and Aquila in the late 1990s when Enron found insurance companies unwilling to insure the company against non-extreme weather events such as the company experienced during a period of mild US winters from 1997 to 1999 (WRMA n.d. (a); Dischel 2002, p. 3). The deregulation of the US energy market had resulted in lower, more competitive energy prices in the US, a situation which aggravated the problem by restricting energy suppliers' ability to extract a surplus from consumers in order to cover periods of unexpected weather conditions (Dischel 2002, p. 3). In order to overcome this barrier, Enron created its own financial product—the weather derivative—taking inspiration from the energy futures markets in which it was involved. The development of the product as a derivative (and therefore a financial, rather than insurance, product) allowed Enron to avoid the regulatory constraints placed on energy companies' use of insurance products (Randalls 2010).

While weather derivative contracts are traded across all forms of weather event, by far the most popular contracts have been based on temperature and the divergence of the average daily temperature from 18 °C. These products, which are popular with firms in the energy industry, are known as Heating and Cooling Degree Days (HDD and CDD) contracts (WRMA n.d. (b)). Over recent years, however, the primary market which provides derivative contracts to end-user businesses has diversified, and a wider and more complex range of products are being developed across a range of weather conditions. One such product is the quantity-adjusting option, or quanto, derivative which combines weather and commodity price risk within a single derivative contract. For example, a company could receive a pay-out on a contract if the temperature is lower than expected, but the pay-out would be calculated in relation to the price of gas (Risk.net 2010).

Such developments illustrate new innovations in the weather derivatives market, however overall the success of the market over the last two decades has been mixed. The weather derivatives market saw massive growth in the mid-2000s, experiencing both the hedge fund boom of 2005–2006 (notional trading value of $45 billion) and the pre-crash boom of 2007–2008 ($32 billion) (Randalls 2010). As with other forms of financial product, the vulnerability of the weather derivatives market was highlighted when the market crashed during 2008–2009 and 2009–2010, with only slow signs of growth by 2011 ($11.8 billion) (WRMA 2009, 2011). These figures, based upon surveys undertaken by PricewaterhouseCoopers on behalf of the WRMA, cover the period 2003–2011. No surveys have been published since 2011, and no up-to-date figures for the size of the market therefore exist. However, despite the dip in the market, in 2011 the WRMA (2011) was hopeful for weather derivatives, pointing to continuing growth outside the US markets throughout the downturn, growing interest in non-temperature-related weather derivatives, and increasing interest from outside the energy industry, and more recent industry reports suggest the market is beginning to expand again (Thind 2014).

## GETTING THE DATA TO MARKET

As financial products based on vast indexes of weather observations, traders in the weather derivatives markets require access to significant amounts of historical and real-time meteorological data. While there has been significant diversification in the sources of data used by market actors in recent years, data produced by national meteorological agencies is still perceived to be preferable due to its high quality and public agencies' substantial archives of historic data.

In the early days of weather derivatives trading, the right to re-use without charge weather data produced by national meteorological agencies was a prominent discourse at industry events. Over recent years, focus on this issue has reduced, however the ease with which market actors can access and re-use publicly funded meteorological data is still perceived to be a significant issue, and a lack of freely available data in some countries is perceived to be a barrier to market growth. In particular, it has been widely noted that while in the US weather data has been in the public domain and freely available for anyone to re-use since before the development of weather derivative markets, in some competing markets, such as the UK, the large volumes of weather data required by the climate risk industry have been treated as a commodity to be traded by the national meteorological agency (Weiss 2002).

Those promoting the development of weather derivative markets in the UK, including lobbyists for the UK financial services industry, such as Lighthill Risk Network (of which Lloyds of London are a member), have spoken out against this practice for a number of years (Department for Business Enterprise and Regulatory Reform 2008). They have called for Met Office data to be made available to commercial users at marginal cost (which tends to be zero for digital resources), so that traders can freely access and use it and, therefore, compete more effectively with the US markets. In the early days of the markets, for example, Weiss (2002) observed that limited access to weather data in the EU had, by 2002, resulted in a weather and climate risk management industry 13.5 times smaller than the nascent US industry, which by this date had built up US \$9.7 billion dollars of contract value over five years.

These demands have filtered down into UK government policy-making. For example, policy documentation developed by senior policy makers in what was at the time named the Department for Business Enterprise and Regulatory Reform (2008, p. 52) indicates support for the financial industry's demand for free use of weather data in order to boost the UK's weather derivatives market. While policy developments began slowly, the election of the new coalition government in the UK in May 2010 led to the demand for free use of meteorological data being quickly incorporated into the government's flagship Transparency and Open Government Data agenda, in line with Open Data advocates' campaign against the commercialisation of public sector data. In the Autumn Statement of 2011, the policy developments came to a head with the announcement by Chancellor of the Exchequer George Osborne, that the UK government was opening "the largest volume of high quality weather data and information made available by a national meteorological organisation anywhere in the world" for anyone to re-use without charge (HM Government 2011). Senior politicians advocating for Open Data, such as MP Francis Maude (2012), spoke publicly about the advantages for the climate risk industry and, according to well-placed policy-makers interviewed by the author, some hoped these developments would make the UK weather derivative market competitive with the US-based markets (Bates and Goodale 2017). Yet, despite the hopes of these key political actors, opening the UK's meteorological data has faced challenges as the Met Office has struggled to adapt to a fully open data environment (Bates and Goodale 2017). Nevertheless, despite these barriers, the climate risk industry is still able to access and process vast amounts of meteorological data in order to trade weather derivate and related climate risk products—albeit with some charges still in place.

## Hedging Against the Climate and the Exploitation of Uncertainty

For advocates of these climate risk products, one of the key benefits is considered to be that they reduce firms' exposure to financial volatility resulting from climate instabilities. While in the long-term a business should expect to pay more into climate risk products than they receive in pay-outs, the business should also expect gains due to having a less volatile profit margin (Dutton 2002, p. 208). For example, the business should be better able to secure credit and protect its market value. For this reason, many perceive that climate risk products increase the "resilience" of businesses and other end users as they adapt to climate change (Michel-kerjan 2013), allowing them to effectively "eliminate the effects of weather and climate from the income statement" (Dutton 2002, p. 209). As Dischel (2002, p. 19) states:

> The goal of hedging is to be less concerned, or not concerned at all, about the impact of weather on cashflow or return. Management achieves freedom from the weather when it engages in a hedge.

At the same time as increasing the 'resilience' of industries and countries that are vulnerable to climate change, climate risk products, it is claimed, offer a substantial growth opportunity for markets to take advantage of in the coming years. For some, therefore, climate risk products are seen as a double win: helping to stabilise economies as firms navigate the uncertain weather conditions that climate change brings *and* simultaneously making substantial profits that contribute to overall economic growth, particularly in the financial centres of the global economy.

Many liberal economists would argue with Stiglitz (2012, pp. 42–43) that within a capitalist economy market failure occurs when there is either imperfect competition, externalities (when a group is affected—positively or negatively—by others' economic activity), information asymmetry or when risk markets are absent. Some might therefore argue that the development of climate risk markets and the increasing availability of free meteorological data might counter some of the market failure problems posed by climate change.

However, in relation to the mitigation of climate change there are risks in the development of weather derivative markets that could lead to negative societal outcomes. This is because they enable the finance industry to generate substantial profits from products that aim to create a sense of security for end-users concerned about climate instability, allowing them to be "less concerned, or not concerned at all, about the impact of weather on cashflow or return" (Dischel 2002, p. 19). These climate risk markets in effect risk reducing the incentive of powerful economic actors to take and demand significant action to mitigate climate change. Such a scenario could increase the negative impact of the actions of those benefiting from these markets, at the expense of the majority, particularly those most vulnerable to climate instability. As more speculators enter the market (Thind 2014), the increase in climatic uncertainty as mitigation efforts fail also presents a growing opportunity to extract profit from the crisis.

While climate risk markets have many interested parties advocating on their behalf, there are others that are more sceptical. Melinda Cooper (2010), for example, argues that weather derivatives are "a claim over the future in all its unknowability—a claim over event worlds that have yet to actualize in space and time" (p. 181). In her analysis of some of these deep seated uncertainties, Cooper (2010) draws on documents produced in 2008 by the US Government's National Intelligence Council and the US non-profit Center for a New American Security to argue that, in the world of US strategic scenario planning, "turbulence" in relation to financial markets, climate change, and energy (p. 169) is no longer perceived as something that there is a possibility of managing and avoiding; rather, "turbulence…is assumed" (p. 184). She argues that, as US strategists have attempted to understand what these deepening uncertainties mean for US geopolitical power in the context of shifting economic power, they have turned to "turbulence" as a form of "productivity" (p. 170) to be leveraged in order to achieve the key strategic aim of sustaining US geopolitical power. One critical objective of US strategists aiming to navigate through these uncertain waters, she observes, is "to dominate… the securitized risk markets, in which weather turbulence plays an increasingly significant role [and which]…offer one possible exit strategy from the liabilities of the dollar-oil nexus" (p. 170).

To turn to the anthropogenic (human-generated) turbulence and uncertainty that is the outcome of the last few hundred years of economic development as a source of new economic opportunity and growth betrays a rigidity in thinking about how societies might move beyond the challenges posed by carbon capitalism. While the development of new data analytics techniques aimed at exploiting turbulence may buy time in the short-medium term, climate risk and similar markets do not address the fundamental problem of economic growth and ecological sustainability in the long term. Drawing upon the analysis of Rockström et al. (2009b), political economists Hay and Payne (2013) argue that many of the earth's ecosystems are already in the 'red zone' and further degradation is directly related to "aggregate global [economic] growth rates". This leads them to conclude that "we face not just a crisis of growth, but, much more significantly, a crisis *for* growth…we will need to wean ourselves off growth if we are to do anything that takes us out of the 'red zone'" (p. 6).

Their argument that economies must move beyond growth in order to shift towards a sustainable form of development is not new, and has been approached from a number of perspectives. The Club of Rome's 1972 report on the Limits to Growth (Meadows et al. 1972), for example, was the first analysis of the problematic relationship between economic growth and environmental sustainability, and these ideas have been taken up across a range of fields including sustainable development and ecological economics. Further, as Harvey (2011) argues, continued economic growth is not only environmentally unsustainable, but is implausible in the long-term given that the rate of profitable investments to be found each year in order to maintain the compound growth that capitalism is dependent upon will at some point become impossible to maintain.

Despite the observed efforts to leverage big data and new forms of data analytics in order to extract economic value from climatic uncertainty, the above arguments suggest that business as usual appears to no longer be an option in the longer term. Of course, while the argument for moving beyond growth as a measure of economic success might be relatively easy to conclude at the abstract level, the actual process of transition towards a new mode of economic development is more fraught and, of course, deeply political.

## Conclusions

It is evident that various forms of power—including informational power—are being deployed in the development and promotion of climate risk products which, in part, aim to respond to the uncertainties posed by climate change. It is also apparent that these products benefit established interests, while, perhaps in some cases unintentionally, deepening the threats faced by the majority, particularly the most vulnerable in society. In the case of the UK government's efforts to open significant amounts of public meteorological data in an effort to leverage the development of the UK's climate risk market, we can observe an example of data policy being used by a government in an attempt to promote a deeply neoliberal, market-driven response to the conditions of uncertainty that we are facing in the early twenty first century.

These developments in data analytics and policy are being shaped to enable particular forms of response to conditions of uncertainty. Still, it would be problematic to argue that increased rights to access and use weather data, and the development of new data analytics techniques are something to be resisted—information is, after all, also necessary for those seeking to establish sustainable, democratic and ecologically sound political economies. However, significant questions remain unanswered about the societal and ecological impacts of climate risk markets, particularly in relation to their potential to disincentivise economically powerful actors' engagement in climate change mitigation activity, and the socio-economic implications of empowering financial elites' efforts to exploit deepening climate uncertainty. There is therefore a need to open up a debate about these forms of financialisation and think critically about how they might, and might not, impact wider efforts to respond to climate change and build more equal and inclusive societies.

# CHAPTER 8

# The Next Internet

## Vincent Mosco

Following from his 2014 book, *To the Cloud: Big Data in a Turbulent World,* Vincent Mosco sees the internet of things, cloud computing and big data analytics as the three pillars of the Next Internet. The Next Internet will be more highly centralised, commercial and controlled than the version we are used to. Mosco posits that it has great potential for social good, but also poses significant social problems.

It would be presumptuous to map out the precise composition of the next stage in the digital world, but it is reasonable to conclude that the Internet, as we have known it for almost three decades, is changing and that the Next Internet may do more to disrupt the world than its older sibling. The Next Internet is far from fully formed and still bears some of the characteristics of the original one born in 1989. But it is growing rapidly and is already challenging its founders' vision of a democratic, decentralized and pluralistic digital world. The Next Internet brings together three interconnected systems: cloud computing, big data analytics and the Internet of Things. It promises centralized data storage and services in vast digital factories that process massive streams of information gathered by networked sensors stored in every possible consumer, industrial and office device, as well as in living bodies. But the Next Internet is

V. Mosco (✉)
Queen's University, Kingston, Canada
e-mail: moscov@mac.com

© The Author(s) 2017

95

B. Brevini and G. Murdock (eds.), *Carbon Capitalism and Communication,*
Palgrave Studies in Media and Environmental Communication,
DOI 10.1007/978-3-319-57876-7_8

also creating major environmental, privacy and labour challenges. The change is so massive and the upheaval so sudden that it is forcing leaders to quickly rethink the models that have described our dominant system of communication, prompting even some technology executives to consider the concept of an information utility. As one CEO puts it, "In the not too distant future, cloud computing will become a 'dispersed utility' and we will come to regard it in much the same way that we view our other core utilities such as gas, water and electricity" (Bridgwater 2016).

The brilliance of the original Internet was figuring out how to get a decentralized, distributed world of servers to talk to one another and thereby connect users through simple, universal software standards. This began to change with the growth of cloud computing, symbolized best by the enormous data centres that have sprung up, seemingly overnight, all over the world. The cloud is a system for storing, processing, and distributing data, applications, and software using remote computers that provide information technology (IT) services on demand for a fee. Familiar examples include Google's Gmail, Apple's iCloud and Microsoft Office, which increasingly distributes its widely-used word processing and business software through the cloud for a monthly fee.

The cloud enables businesses, government agencies and individuals to move their data from onsite IT departments and personal computers to large data centres located all over the world. What is saved in storage space also opens a rapidly growing business for companies that profit from storage fees, from services provided online and from the sale of customer data to firms interested in marketing products and services. Government surveillance authorities like the National Security Agency (NSA) and Central Intelligence Agency (CIA) also work closely with cloud companies, particularly Amazon, to meet their security and intelligence needs (Kunkel 2014). The diverse collection of servers providing the foundation for the original Internet has evolved into a centralized, global system of data centres, each containing tens or hundreds of thousands of linked servers, connected to the world through telecommunications systems, and operated primarily by private corporations and government military and surveillance agencies. The leading science journal *Nature* made very clear the practical difference between the original Internet and one based in the cloud when it called on the US government to establish a "Cloud Commons", one version of an information utility, for biological research, especially in genomics. It did so because research on large data sets is far easier and faster to carry out in the cloud than through servers based in university research facilities (a difference in project time alone of between 6 weeks for the cloud and 6 months for the old Internet) (Stein et al. 2015).

The cloud is more like a data factory than a storage warehouse because it processes data to produce services, such as marketing, accounting, customer relations as well as legal and financial services. That makes companies and government agencies partners in service provision with the companies that own and manage data centres. It also marks a major step toward creating a centralized, globalized and fully commercial Internet that resembles giant water and electric utilities. The major cloud providers are almost all large corporations led by Amazon, by far the world's largest cloud business. It is trailed by Microsoft, IBM and Google. Through service contracts, most of these are well integrated into the military, intelligence and surveillance arms of government. Amazon, for example, provides cloud computing storage and services for both the CIA (through a $600 million contract) and the NSA. Meanwhile government agencies demanding heightened levels of security are building their own cloud facilities, including the NSA, which in 2015 opened one of the world's largest, in a remote mountain location in Utah.

Big data analytics makes up the second leg of the Next Internet. In spite of the proliferation of fancy new titles, like data science professional, that fuel enthusiasm, there is very little that a social scientist would find novel in the big data approach. It generally involves taking a large, often massive and almost always quantitative data set, and examining the specific ways the data do or do not cohere or correlate, in order to draw conclusions about current behaviour and attitudes and go on to make predictions. The aim is to produce algorithms or a set of rules that specify conclusions to be drawn or actions to be taken under specific conditions.

Facebook, for example, takes the data generated by its 1.7 billion or so users and relates the 'likes' associated with posts about everything from celebrities, companies and politicians to views about society, products and, of course, cats. These enable the company to develop profiles on its subscribers, which Facebook sells to marketers who target users with customized ads sent to their Facebook pages, what years ago in the pre-social media age, Gandy (1993) called the panoptic sort. Google does the same for search topics as well as for the content of Gmail, and Amazon creates profiles of its users based on searches and purchases on its site. Given the limitations of quantitative correlational analysis, especially the absence of historical context, theory and subjectivity (qualitative data is ignored or poorly translated into numbers), such analysis is not always accurate and incidents of big data failures, on such projects as seasonal flu forecasting and building models for economic development, are mounting, as are the opportunities to make mischief with data for profit (Mosco 2014). One has to look no further than the 2016 US presidential election when big data analysis not only failed to forecast the outcome. It may also have shaped the result because it produced flawed algorithms that led the Hillary Clinton camp to take an overly cautious approach because it believed data suggesting she was the clear leader and likely winner, what amounted to a massive case of what Gandy would appreciate as a panoptic missort. Nevertheless, for simple questions such as what are the likes and dislikes of every conceivable demographic cohort or for drawing conclusions about users based on their friendship and follower networks, the massively large stores of data available for analysis in the digital factories that make up the cloud, offer major incentives for companies and governments to invest in data centres and in big data analysis. It is reasonable to be concerned that singular reliance on big data in research is paving the way for what might best be called *digital positivism*, a methodological essentialism that ignores history, theory and subjectivity.

The cloud and big data are enhanced substantially by the growth of the Internet of Things. From watches that monitor blood pressure to refrigerators that order fresh milk, from assembly lines "manned" by robots to drones that deliver weapons, it promises a profound impact on individuals and society. The Internet of Things refers to a system that installs sensors and processing devices into everyday objects (watches), production tools (robotic arms) and armaments (weaponised drones) and connects them in networks that gather and use data on their performance. The sensors in a refrigerator form a network of things that report on what's inside and how it is used. The Internet of Things is made possible by advances in the ability to miniaturize scanning devices and provide them with sufficient processing power to monitor activity, analyze usage and deliver results over electronic networks (Greengard 2015).

A 2015 report from the private think tank McKinsey concluded that by 2025 the Internet of Things will have an economic impact of between $3.9 and $11.1 trillion (US) which, at the high end, is over 10% of the world economy (Manyika et al. 2015). Significantly, it is the manufacturing sector, and especially General Electric, that leads the way as machine production and opportunities for operational surveillance enable more tightly managed and efficient factories and global supply chains. But these will also extend, McKinsey maintains, to offices, retail operations, the management of cities and overall transportation, as automated vehicles take to the streets and highways made 'smart' by sensors embedded everywhere. Heightened monitoring will also extend to the home, promising greater control over heating and cooling, ordering food and supplies and to the body as well, where sensors will continuously monitor fitness, blood pressure, heart rate and the performance of vital organs. This sounds futuristic and, depending on your point of view, either dystopian or utopian, but it speaks to the power of the new technology and to the fundamental differences between the original Internet and its successor.

Companies have been quick to take advantage of their leading positions in the digital world to expand into the Internet of Things. Prime examples include Google's driverless car, the Apple Watch and Amazon's embrace of robotics in its warehouses to speed the work of order fulfillment. Amazon is also beginning to use drones for deliveries, and is developing entirely new forms of packaging containing pushbuttons that automate ordering refills. The Internet of Things has also given new life to an old industrial firm, General Electric, which was remade in the 1990s by shifting from manufacturing to finance. GE has now all but abandoned the increasingly regulated world of banking only to emerge as a dominant player producing devices essential to the Internet of Things and making use of them in its own industrial processes. Along with the benefits to corporations, the Internet of Things holds out great promise for the military, because it greatly strengthens opportunities to automate warfare through robotics and drone weapon delivery, in addition to enhancing and automating the management of troops (Gusterson 2015).

For business, one enormously valuable result of monitoring every device and connecting them in a global grid of objects is the exponential growth in commercially useful data. Making use of this surge in data will require both new cloud data centres and widespread use of data analysis. As Manyika et al. (2015) put it, "Currently, most Internet of Things data are not used. For example, on an oil rig that has 30,000 sensors, only one percent of the data are examined. That's because this information is used mostly to detect and control anomalies—not for optimization and prediction, which provide the greatest value." How to use data, internally and as a marketable commodity, is one of the biggest challenges facing the Internet of Things industry.

Most of what is written about the Next Internet is technical or promotional, emphasizing the engineering required to build it or touting the potential in sometimes dreamily hyperbolic terms: nonstop leisure, friction-free capitalism and the Singularity. We are just beginning to see some discussion of the serious policy issues that arise in a world of massive data centres, nonstop analysis of human behaviour and ubiquitous connectivity. These include the concentration of power over the Next Internet in a handful of mainly US companies and the military-intelligence apparatus; the environmental consequences of building and maintaining massive data centres and powering systems, threats to privacy and security and the impact of automated systems on human labour.

Two things stand out about the early configuration of the Next Internet industry. It is already highly concentrated and is dominated by American firms. Indeed, on August 1, 2016, the top five Next Internet companies were also the world's top five in market value. These include Amazon, which controls over one-third of the cloud computing market and has a formidable presence in big data and the Internet of Things. The company was among the first to build a one-size-fits-all cloud service that attracted individuals and organizations with its simplicity and massively discounted prices that suggest the not-so-fine art of predatory pricing was at work here. Google, Microsoft, Facebook and Apple round out the list of firms that use their control over the original Internet to become leaders in the Next Internet. Legacy companies like IBM, Oracle, HP and Cisco have scrambled to replace their expertise in servicing IT departments that are now disappearing and pivot to the new digital world. However the need to cannibalize old systems and remake their organizations has made the going slow. In addition, there are firms that specialize in one or another of the constituent Next Internet systems, such as Rackspace and Salesforce.com, but these are constantly undermined by encroachment from the dominant companies. A force of potentially great significance in the Next Internet arena is General Electric, which is betting heavily on reinventing factories with the Internet of Things.

Historians of technology will recognize the similarity of this pattern of ownership concentration to the early days of electrification, telegraphy, telephony and broadcasting. In each of these cases, regulation and outright state ownership were required to control commercial abuses and increase access at affordable rates. However, these remedies are less likely to be applied in a world where regulation and government ownership are no longer in favour. Moreover, as in the past, dominant firms are benefitting from their close ties to the military and intelligence agencies, providing them with Next Internet services and cooperating, more often than not, with requests for information on users. In fact, close ties to the Pentagon, including its well-funded research arm the Defense Advanced Research Products Agency (DARPA), as well as to the NSA and the CIA, helps to explain why there are no challengers to US hegemony over the Next Internet coming from Europe, whose telecommunications companies once led the world.

China provides the only serious competition. There, government has invested heavily in Next Internet technologies going as far as to integrate them into its five-year plans and build entire cloud cities. This has benefited leading companies like Alibaba, Baidu, Huawei, and Tencent, among others. Signalling that it intends to challenge America's lead, Alibaba has set up shop in Silicon Valley and, like other Chinese firms, is building on the enormous domestic market to extend its reach internationally (Tse and Hendrichs 2016). A look at the remaining policy issues reveals why the concentration of corporate power is such a significant problem and why it is essential that societies begin to consider the need for public intervention to regulate and control the Next Internet as an information utility.

Because the digital world is made up of invisible electrons zipping through the air, there is a tendency to view it as immaterial. Nothing could be further from the truth and the sooner this is recognized, the more likely the environmental problems associated with the Next Internet will be addressed. Cloud data centres are very material structures and, as they come to fill the world, there are numerous emerging environmental policy issues. It is expected that by 2017 data centres will consume 12% of the global electricity grid (Sullivan 2015). Moreover, customer demand for 24/7 services requires several layers of backup power, including lead acid batteries and diesel generators that have been found to be carcinogenic. Furthermore, many data centres require large, continuous supplies of water for their cooling systems and this raises serious policy issues in places like the US West, where years of drought have taken their toll. So far, data centre operators have used their economic power and the allure of promised jobs to successfully pressure local governments to provide property tax breaks, cut-rate power deals, and relief from pollution regulations.

Some companies have responded to opposition from environmental groups, especially Greenpeace, by incorporating solar and other sustainable energy sources into their data centre power supplies. But as data requirements grow, systematic regulation is required, including a broad review of discount power deals, the use of massively polluting backup systems, and the diversion of water resources to cool servers. Notwithstanding any progress in this area, the primary source of power consumption in the Next

Internet is from the sensors embedded in the billions of connected devices and from the communication systems that link people and things through cellular and other wireless networks. A world of ubiquitous, always-on connected devices is enough to make energy executives salivate, especially the lobbying arm of the coal industry which views the Next Internet as an opportunity to build on what a study for the US National Academy of Sciences calls "the renaissance of coal" (Steckel et al. 2015). As a report sponsored by the coal industry concluded, "The inherent nature of the mobile Internet, a key feature of the emergent Cloud architecture, requires far more energy than do wired networks.... Trends now promise faster, not slower, growth in ICT energy use" (Mills 2013). When the environmental impacts of Next Internet systems are considered alongside their massive stimulation of consumption, the implications for climate change are staggering.

Privacy and security concerns rise exponentially in the Next Internet because greater connectivity increases opportunities for technical breakdowns and criminal hacking. Indeed one tech journalist referred to the Internet of Things as "the greatest mass surveillance infrastructure ever" (Powles 2015). By the standards anticipated in a digital world where the Internet of Things is fully developed, today's Internet is far from creating a connected world. About 40% of the world's population now uses the Internet at least once a year, and, as one might expect, access is concentrated in the developed world and in urban centres (Gagliordi 2015). With only 1% connectivity among objects, we are far from the vision of ubiquitous computing. But even at this relatively low level, technical problems and criminal hacking plague the system. On one day alone in 2015, the entire US fleet of United Airlines planes was grounded, the New York Stock Exchange shut down for several hours and the Wall Street Journal's computers simply stopped operating. All of these were explained as the result of technical 'glitches'. Just as this calamity hit the news stream, the US government reported that hackers had stolen the personnel records of 22.1 million federal employees, contractors, and their families and friends who provided information for background checks. The haul also included over one million sets of fingerprints (Nakashima 2015). In the largest reported hack of all, in December 2016, Yahoo reported that hackers had made off with the records of one billion people in 2013 and were selling the data on the Dark Web, the corner of the Internet inhabited by criminals, spies and those seeking stolen or illegal goods (Goel and Perlroth 2016).

Hacking aside, the most significant threats arise from data-hungry businesses and governments. For them, the greatest attraction of ubiquitous computing is the valuable data on the behaviour of people and the performance of objects. These offer companies the opportunity to refine targeted advertising and product development well beyond the crude systems that today's Internet makes possible and they enable governments to deepen their ability to track and control citizen behaviour and attitudes. Consider the commercial benefits to insurance companies that will be able to continuously monitor the health of customers, their driving habits, and the state of their homes; or to governments that can adjust benefits and other services based on citizen behaviour registered in their actions, as well as their interactions with one another and with the things that fill their lives; or to employers that are even now requiring office workers to wear sensor devices on and under the skin for ubiquitous performance monitoring (Wilson 2013). Discussions of anticipatory selling as well as of algorithmic policing, euphemistically called 'predictive analytics', are worrisome to privacy advocates because they are attracting great interest from businesses and governments (Davenport 2014).

The impact of the Next Internet on jobs and the nature of labour is also an important policy issue. At first glance, it is tempting to think 'here we go again' because the impact of technology on jobs has been discussed for many years but especially since the end of World War II when computer scientist Norbert Wiener generated considerable public debate by raising the spectre of massive job loss due to automation (Wiener 1948). Moreover, the Next Internet is creating and will likely continue to create employment, including traditional construction jobs in the build out of global networks of data centres, in the new profession of data science, and in the control, maintenance, and monitoring of networked things. There is another reason why it is important to approach the impact of computer technology on jobs and the economy with caution. As research documents, overall employment has been much more closely tied to gross domestic product (GDP) than to computerisation and, except for the late 1990s when there was massive investment in hardware, the long-promised productivity gains from IT have failed to materialize (Gordon 2016).

However, today there are far more opportunities for the new technology to eliminate human labour, especially professional knowledge work. In fact, one expert consultant prefers to define cloud computing as "nothing more than the next step in outsourcing your IT operations" (McKendrick 2013). This is in keeping with a general tendency, which one researcher for

Gartner Associates summarizes succinctly: "The long run value proposition of IT is not to support the human workforce—it is to replace it" (Dignan 2011a). The Next Internet creates immediate opportunities for companies to rationalize their information technology operations. Again, from Gartner, "CIOs believe that their data centres, servers, desktop and business applications are grossly inefficient and must be rationalized over the next ten years. We believe that the people associated with these inefficient assets will also be rationalized in significant numbers along the way" (Dignan 2011a).

Next Internet companies maintain that their systems can break a pattern in business organizations that began when the first large computers entered the workplace. Back then all business and government agencies insisted that it was essential to operate their own IT departments and, for larger organizations, their own data centres. Next Internet supporters insist that it is no longer essential to build and run thousands of organization-specific facilities when a few large data centres can meet the demand at lower cost with far fewer professional personnel. This process has already begun and early studies demonstrate that, even with limited downsizing of IT departments, companies are saving between 15 and 20% of their IT budgets (Howlett 2014).

The Next Internet also makes possible the widespread rationalization of practically all knowledge and creative labour because the work of these occupations increasingly involves the production, processing and distribution of information. According to one observer, "In the next 40 years analytics systems will replace much of what the knowledge worker does today" (Dignan 2011b). A 2013 report concluded that almost half the current US workforce is directly threatened and in the high-risk category for job loss (Frey and Osborne 2013). Whatever the precise share, there is no doubt that the current trend is to use software to move knowledge worker labour to machine systems. We are now beginning to see the impacts on education, health care, the law, accounting, finance, sales and the media. Private and public sector organizations are encouraged to outsource all but their core business processes to companies like Salesforce.com, which specializes in managing vast databases of customer information, a job that marketing and client service departments inside companies typically performed.

The expansion of outsourcing to computers raises serious questions for the entire global system of flexible production. According to Gartner, "That outcome will hit all economies—especially emerging ones like India that now dominate technology outsourcing" (Dignan 2011a). The Next Internet also expands the range of potential outsourcing practices. It may be an overstatement to declare, as did *Forbes* magazine, "We are all outsourcers now," but it certainly makes feasible more kinds: "Outsourcing is no longer simply defined by multi-million-dollar mega-deals in which IT department operations are turned over to a third party. Rather, bits and pieces of a lot of smaller things are gradually being turned over to outside entities" (McKendrick 2014). Amazon is a leading force in this process with its Mechanical Turk business that charges individuals and organizations to outsource micro-tasks to a worldwide reserve army of online piece workers. Combined with the promise of product warehouses full of robots to locate, pack, and ship goods and drones to deliver them, Amazon is the leading edge of the Next Internet's push to expand labour intensification throughout the world. Whatever the impact on the number of jobs, the Next Internet is already changing the labour process. Workers at a Swedish firm can attest to this as they arrive at the office each day with RFID chips implanted under the skin to improve productivity and management control (Cellan-Jones 2015).

What can be done to address these problems? First and foremost, it is essential to view them as intrinsically social and not just technological. While technology plays a role in addressing serious policy issues, there is no simple digital fix to solve them. It will take concerted political action to tame the concentrated corporate power that is now making the Next Internet a tool to expand the power and profit of a handful of digital giants. It will also take global social movements, stronger versions of what supporters called a New World Information and Communication Order in the twentieth century, to build a digital commons for the twenty-first. Furthermore, we need to make environmental protection and sustainability central to all decision-making about the Next Internet. It is also important to rethink privacy as the human right of access to the psychological space essential to develop individual autonomy. Above all, privacy is an essential right of citizenship and not a tradable commodity. Protection of personal, interpersonal and autonomous space from commercial and government surveillance must also be central to the choices made about the Next Internet. Finally, we need social policies about employment and income that address the state of human labour in an age when automation threatens jobs, including now those of the white-collar workforce, and massive invasive surveillance threatens fundamental worker rights. Does this mean we should reopen the discussion of a guaranteed annual income?

What *is* the right balance between job creation and such a guaranteed income? How can we facilitate organizing digital workers who tend to be employed in the "gig" economy of precarious jobs? Are unions at Salon and Vice, and worker associations at Uber and Lyft, all pioneering web-based successes, good models for the future?

The digital world is at a critical juncture represented by two clashing visions. The first imagines a democratic society where information is fully accessible to all citizens as an essential service. In this view information is managed through forms of regulation and control that are governed by representative institutions whose goal is the fullest possible access and control for the greatest number of citizens. Governance might take multiple forms, including different combinations of centralized and decentralized approaches at local, regional, national and international levels. The second envisions a world controlled by global corporations and the surveillance and intelligence arms of national governments. Under this model, the market is the leading force shaping decisions about the production, distribution and exchange of information and corporations with market power hold the most influence. In this fundamentally undemocratic world, digital behemoths share power with governments that make full use of technology for surveillance, control and coercion.

Fifty years ago, long before the first Internet, the Canadian scholar and policy analyst Douglas Parkhill chose the democratic vision in his book about the need to create a global system of computer utilities that would guarantee public control and universal access. Social movements had helped to tame private monopoly power over essential resources like water and electricity by making them public utilities. Parkhill (1966) made the case that information was no less essential and no less in need of public control. The Next Internet is an opportunity to build on this vision.

The utility concept received a boost when the US Federal Communication Commission issued a 2015 ruling affirming the right of people to fair and equal access to the Internet, what is popularly known as the net neutrality decision. Although limited, the decision sparked hopes that we would begin to see the Internet not as the property of commercial and military interests but as a public commons, controlled by citizens. In its call for "platform cooperatives" Jeremy Corbyn's 2016 "Digital Democracy Manifesto" proposes a rich, contemporary version of the public utility idea that has received support from progressive Internet and social movement activists (Scholz and Schneider 2016). The rise of cloud computing, with its extreme centralizing tendencies, has brought about renewed interest in the utility concept including among technologists and entrepreneurs. Given the massive environmental, privacy, and labour issues associated with the Next Internet and the abject failure of existing policy processes to deal with them, it is imperative to consider how to create democratic and publicly controlled communication networks and the political system that would use them in the public interest.

# PART III

# Chapter Seven

# Corporate Capture: PR Strategies and Promotional Gambits

**Fig. III.1** Here to Debunk

# An Interview with Kim Sheehan: Greenwashing in the Experience of the Greenwashing Index

## *Kim Sheehan and Benedetta Brevini*

Bringing more than 12 years of experience in advertising and marketing to the University of Oregon's School of Journalism and Communication, Kim Sheehan has consulted for numerous companies including *People magazine* and Laura Ashley. She is a past president of the American Academy of Advertising.

### GREENWASHING

**BB: Greenwashing has been defined as a form of spin in which green marketing and PR strategies are used to promote the perception that an organisation's practices are environmentally friendly. Would you agree with this definition and can you give us an example?**

K. Sheehan (✉)
University of Oregon, Eugene, USA
e-mail: ksheehan@uoregon.edu

B. Brevini
University of Sydney, Sydney, Australia
e-mail: benedetta.brevini@sydney.edu.au

© The Author(s) 2017
B. Brevini and G. Murdock (eds.), *Carbon Capitalism and Communication*,
Palgrave Studies in Media and Environmental Communication,
DOI 10.1007/978-3-319-57876-7_9

KS: I sort of agree with this. I would add to "promote the perceptions that an organization's practices are more environmentally friendly than they really are". Your definition would have any green message as greenwashing, and that isn't really true.

An example, and where greenwashing started, is the card in a hotel room that tells you that not getting clean linens every day is good for the environment. Yes, it is, but it is better for the hotel because they save money in labour, water, energy and equipment costs. And the horrible cleaners they use to clean rooms are horrible for us and the environment.

**BB: Can you give us an idea of how big this phenomenon is and how much it increased in the last decade?**

KS: Green advertising has peaks and valleys. When the economy is good, green advertising increases. When the economy is bad, green advertising decreases. Clorox has slashed its ad budget for its Greenworks line, and they aren't the only company that has done so. Part of the reason why is the increased government scrutiny into green messages, begun with the implementation of the newest set of [Federal Trade Commission] Green Guides.

So I don't think we're necessarily seeing an increase in green messages, or in greenwashing. I think many green messages can be considered greenwashed.

**BB: Do you think that commercial and mainstream media have had a prominent role in amplifying greenwashing practices?**

KS: Not really. I think that there are more investigations on why things aren't green than promoting green products that aren't really green.

## THE INDEX

**BB: How did the idea of a Greenwashing Index come about? When was it launched? What are the purposes?**

KS: We wanted to provide a place with green advertising could be scrutinized and celebrated (when deserved) and not celebrated (when deserved). We also wanted to be able to understand what consumers want and need from green messages in order to help inform the development of the FTC Green guides, which was happening in 2007.

**BB: How is it currently supported financially?**

KS: The ad agency EnviroMedia supports the site.

**BB: Can you explain in detail how the Greenwashing Index scoring criteria work? From the information submitted by consumers, to the actual website management and index?**

KS: Anyone can submit an ad, anyone can rate an ad.
The criteria are:

1. **THE AD MISLEADS WITH WORDS.**
   Do you believe the ad misleads the viewer/reader about the company's/product's environmental impact through the things it says? Does it seem the words are trying to make you believe there is a green practice when there isn't? Focus on the words only—what do you think the ad is saying?
2. **THE AD MISLEADS WITH VISUALS AND/OR GRAPHICS.**
   Do you think the advertiser has used green or natural images in a way designed to make you think the product/company is more environmentally friendly than it really is?
3. **THE AD MAKES A GREEN CLAIM THAT IS VAGUE OR SEEMINGLY UNPROVABLE.**
   Does the ad claim environmental benefits without sufficiently identifying for you what they are? Has the advertiser provided a source for claims or for more information? Are the claims related to the company/product?
4. **THE AD OVERSTATES OR EXAGGERATES HOW GREEN THE PRODUCT/COMPANY/SERVICE ACTUALLY IS.**
   Do you believe the advertiser is overstating how green the product/company actually is? Are the green claims made by the ad believable? Do you think it's possible for the product/company to do the things depicted/stated?
5. **THE AD LEAVES OUT OR MASKS IMPORTANT INFORMATION, MAKING THE GREEN CLAIM SOUND BETTER THAN IT IS.**
   Do you think the ad exists to divert attention from something else the company does? Do you believe the relevant collateral consequences of the product/service are considered in the ad? Does it seem to you something is missing from the ad?

The scores are averaged together for the final index score.

**BB: What have been the most relevant successes of the Index, and what are the obstacles it faces?**

KS: We used the information to inform the development of the FTC Green Guides, and also to raise awareness of the need to scrutinize constitutive communication elements (colours, images, fonts) in addition to the actual rational or pragmatic message given in a message.

The obstacles are drawing awareness.

**BB: How easy it was/is for you to get coverage of your activities in the mainstream media? What about social media? Has that provided a new platform for publicising your work?**

KS: We got a lot of coverage when it kicked off due to the interactive nature of the site, which was unique at the time. The popularity of the site waxes and wanes with salience of the environment. Given the new administration, it is challenging to anticipate where this all will go.

**BB: With the inauguration of Trump's presidency there is clearly more uncertainty in the US about the possibility for climate action. Do you think that this will have an impact on your activities?**

KS: Yes. Our environment is in danger, and it will take more than a scrutiny of branded messages to enact meaningful change.

# CHAPTER 10

# Fighting for Coal: Public Relations and the Campaigns Against Lower Carbon Pollution Policies in Australia

*David McKnight and Mitchell Hobbs*

## INTRODUCTION

Australia is not only one of the world's biggest exporters of coal; it is also heavily reliant on burning coal to run its economy. It has also been the site of several key battles fought by the global fossil fuel industry to oppose measures designed to reduce carbon emissions. In 2008–2009, the coal industry publicly campaigned against the introduction of a carbon emissions trading scheme by the Labor government of Prime Minister Kevin Rudd. The Carbon Pollution Reduction Scheme (CPRS) was designed to be a 'cap-and-trade' scheme whereby the government sets a limit on the amount of greenhouse gases the nation can release into the atmosphere in a given period, with the government releasing a series of carbon pollution permits that are equal to the emissions cap. Companies could then trade these permits to meet their energy consumption and production needs.

D. McKnight
University of New South Wales, Kensington, Australia
e-mail: d.mcknight@unsw.edu.au

M. Hobbs (✉)
University of Sydney, Sydney, Australia
e-mail: mitchell.hobbs@sydney.edu.au

© The Author(s) 2017                                                   115
B. Brevini and G. Murdock (eds.), *Carbon Capitalism and Communication*,
Palgrave Studies in Media and Environmental Communication,
DOI 10.1007/978-3-319-57876-7_10

Thus market forces limit carbon emissions by providing a cost disincentive for the use of fossil fuels. However, the Rudd government was forced to abandon the CPRS before it could be introduced, after the coal industry developed an under-the-radar public relations campaign in regional areas that influenced community sentiments and emboldened climate change 'sceptics' within the parliamentary Opposition.

Shortly after, in 2011–2012, the coal industry fought the introduction of a 'carbon tax' by Rudd's successor as Prime Minister, Julia Gillard. This carbon-pricing scheme required most businesses that emit over 25,000 tonnes per year of greenhouse gases to purchase permits, called carbon units, from the government. The carbon tax was successfully introduced as the *Clean Energy Act 2011*, with the Gillard government intending that it be replaced by an emission trading scheme (ETS) in 2014–2015, where the available permits will be limited in line with a pollution cap. However, like the Rudd government's ETS, this policy also became the centrepiece of a political attack on the Gillard government and was instrumental in its defeat in 2013 by an opposition leader, Tony Abbott, who argued that "coal is good for humanity" (McCarthy 2014).

In this chapter we examine the public relations strategies and tactics used by the coal industry to defeat legislation intended to reduce Australia's contribution to anthropogenic climate change, especially those policies implemented by the Rudd and Gillard Labor governments (2007–2013). The case studies explored below demonstrate the significant power of the coal industry to shape its operational environment in resource rich countries.

## Using Public Relations to Maintain the Status Quo

In their book *Merchants of Doubt*, Naomi Oreskes and Erik M. Conway (2011) explore the different public relations strategies and tactics developed by the tobacco lobby, which mining and energy conglomerates around the world have subsequently adopted. In December 1953, the presidents of four of America's largest tobacco companies met with John Hill, the founder and CEO of one of America's largest public relations (PR) firms, Hill and Knowlton (Oreskes and Conway 2011, p. 150). Hill helped the tobacco industry to devise and implement a public relations strategy that would seek to undermine the development of a scientific consensus on smoking and cancer.

The overarching communication strategy was to foster 'doubt' by challenging tobacco researchers and their evidence. The tactics employed in support of this strategy included: *ad hominem* attacks against scientists and organizations; rhetorical tactics of blame shifting and misdirection; dubious third-party endorsements and testimonials from 'experts'; publishing pseudo-scientific reports and studies; espionage of oppositional groups; and political donations with *quid pro quo* expectations. Oreskes and Conway (2011) contend that these interventions, which sought to influence debate and sow doubt regarding the consequences of smoking, are now employed by the mining and energy industry to argue that the science of global warming is 'not settled'. As discussed below, many of these tactics have been used to oppose attempts by the Australian government to pass legislation to mitigate the country's greenhouse gas emissions.

The communication tactics employed by the coal industry are part of the 'persuasive tradition' of public relations first developed in the US in the early 20th century. According to Edward Bernays, the self-proclaimed founding father of public relations (Gower 2008, p. 308), communication specialists should use information and persuasion "to engineer public support for an activity, cause, movement, or institution" (Bernays 1955, pp. 3–4). While public relations academics and practitioners acknowledge that persuasion can be manipulative, they also argue that it can by ethically justified. Public relations has often sought to change attitudes and behaviours so that they align with the operational interests of corporations and other organizations; how far these interventions also advance the public interest is open to debate (Pfau and Wan 2006).

Persuasive communication strategies are often used by organizations that face 'adversarial publics'—such an environmental groups—in their operational environment, which makes it a common approach for the coal industry (Miller and Sinclair 2009). Advocates of the persuasive tradition in public relations contend that corporations have an obligation to their shareholders and employees, and should take actions to protect their financial interests and the livelihoods of their stakeholders. In practice this means that public relations practitioners seek to challenge critics of their operations through persuasive counter arguments, and see themselves as advocates or defenders of their organizations much like an attorney in an advocacy legal system (Grunig and Grunig cited in Pfau and Wan 2006, p. 103).

Persuasive communication strategies draw on a repertoire of common tactics that include, as a first line of defense, actions which takes place largely behind closed doors. Lobbying key government members is seen as an essential tactic for explaining the organization's point of view to legislators and initiating two-way dialogues capable of convincing either side of the merits or deficits of a particular policy initiative (Harrison 2011; Heath and Waymer 2011).

However, when these internal communication campaigns fail, the organizations involved will often elect to 'go public' and take their arguments and complaints to the media and the 'broader court of public opinion'. The aim is to influence popular understandings and attitudes in order to apply pressure on specific politicians so that they change their stance on a given issue. 'Going public' is also a targeted strategy in that organizations will run public relations campaigns in strategically significant communities (McKnight and Hobbs 2013, p. 311). Common tactics here include organizing town hall meetings, rallies and protests as well as issuing direct mail letters, flyers and pamphlets. These 'ground campaigns' are often supported by advocacy advertisements in the media and online.

As noted by Herbert Waltzer (1988) in his typology of corporate advertising, advocacy and issues advertisements are an aggressive public relations strategy. They are also mostly a *reactive* response to potential legislative agendas or the advertisements of adversarial publics (Miller and Sinclair 2009; Miller 2012). Their aim is to frame people's understanding of an issue or event, a practice communication scholars refer to as 'second level agenda setting' in that it aims to influence how people think by shaping their conceptual schema used in decoding information (Martinelli 2011, p. 40; Entman 2004).

PR theory also provides *proactive* strategies for controversial organization to employ during times of relative peace with their publics. The most common of these is called 'inoculation public relations' (Pfau and Wan 2006) and aims to 'immunise' an organization and their operations from future challenges from legislators or oppositional groups by bolstering their public image. According to Coombs (1998, 1999a) inoculation PR works on the assumption that corporations can accumulate 'image credits' via the publicity of positive performance, which can in turn offset reputational damage generated by a future crisis or an adversarial communications campaign.

Advocacy or 'issue advertisements' are often used as part of an inoculation strategy, with the goal of presenting the organization as a positive contributor to the economy and society more broadly (Burgoon et al. 1995). 'Good works', such as providing financial assistance and sponsorships to community organizations or charities are often publicised in these advertisements in order to craft a 'halo effect' (Coombs 1999b). Political donations to individual politicians and political parties can also be considered an inoculation public relations tactic in that the goal is to build a surplus of goodwill, or 'rational capital', which makes it difficult for parliamentarians to enact legislation that might have an adverse commercial impact.

Taken together, inoculation and persuasive advocacy strategies amount to a 'carrot and stick' approach that seeks to support the ongoing operations of controversial industries. As will be made clear in the following case studies, the coal industry in Australia has operationalised this 'reward and punishment' approach to great effect, securing the cooperation of parliamentary allies, while punishing those who would seek to legislate against their financial interests. Moreover, industry advocates have not been shy about wielding this form of communication power.

## Case Study #1: The First Emissions Trading Scheme, 2008–2010

During 2008–2009 the newly elected Labor government under Prime Minister Kevin Rudd tried to pass laws for the CPRS. This emissions trading scheme proposed to price carbon emissions created in Australia, a country with the highest per capita emissions in the world. At this point, up to 75% of all electricity used was created from burning coal, making the sector one of the biggest polluters. In 2007, before the CPRS was even introduced, the private electricity generators started issuing warnings of possible blackouts if an emissions trading scheme was introduced. The head of the electricity generators' association stated: 'There will be real problems in maintaining a reliable electricity supply.' According to one account, as the trading scheme developed, the electricity companies 'fought back with everything they had... [and] ran a disciplined and ruthless campaign to undermine public support for the Rudd scheme' (Chubb 2014, pp. 40–41).

In early 2009, the Labor government released draft laws for the trading scheme that were immediately attacked by the peak body representing the mining companies, the Minerals Council of Australia (MCA). The MCA Chairman, Peter Coates, who was also chairman of the coal mining company Xstrata, described the trading scheme as 'crazy stuff' (Hall 2009). But the most powerful opposition came from the Australian Coal Association, which represented a local industry dominated by global players such as BHP Billiton, Rio Tinto, Anglo American and Glencore. Thanks to the dominance of such corporations, 80% of the Australian coal industry is foreign owned (Cleary 2011). The campaign that these corporations ran was organised under the slogan "Cut emissions, not jobs". Its wording suggested that the coal industry supported cuts to emissions but this was highly misleading. The CPRS was designed to cover the emission of methane, a powerful greenhouse gas, from the large number of open cut mines on which Australia's coal export business is based. The intention was to make so-called 'gassy' mines less desirable than ones with less gas emissions. But this aspect of the plan aroused fierce opposition from the coal industry in spite of its purported commitment to 'cut emissions'.

The initial targets of the "Cut emissions, not jobs" campaign were in regional coal mining communities in the state of Queensland, an area rich in coal deposits and the centre of Australia's large export industry. The launch of the campaign by the Coal Association in one major Queensland town 'coincided' with a visit to the same town on the same day by a leading climate denier, Professor Bob Carter (Wratten 2009). Other coal mining regions, such as the Hunter Valley and Illawarra coast, were targeted along with regional areas that hosted aluminium smelters. The campaign's public relations tactics targeted regional print, radio and TV media, and several public rallies were also organised in strategic electorates in order to send a message to local parliamentarians (Grigg 2010a). The message to these areas was that 16 mines would be forced to close and 10,000 jobs would be lost if the CPRS went ahead (Australian Associated Press 2009). While this campaign aroused fears in many regional areas, business commentators were pointing out that investment in the coal industry was booming based on predictions that coal exports would double (Manning 2009).

The persuasive PR campaign by the coal mining industry was a major factor in a policy retreat by the Rudd government in May 2009 that saw promises of greater financial support for business and more modest targets for emissions reduction. While the peak bodies of business welcomed the back down, the Minerals Council continued its opposition, claiming that a revised emissions trading scheme would still cost "thousands of mining jobs and billions of dollars in investment" (Breusch 2009).

The concessions extracted by the coal industry made it harder to pass the trading scheme laws in parliament. In August 2009, parliamentarians representing the Greens declined to support the CPRS ensuring that it was defeated. The government then conducted further negotiations with the conservative opposition led by Malcolm Turnbull. However, these broke down when the conservatives elected a new leader, Tony Abbott, who was on record as claiming that climate change science was 'crap' and was strongly opposed to the ETS (Readfearn 2014). Chief Executive of the Minerals Council of Australia, Mitchell Hooke, would later boast to journalists about his role in changing the leadership of the parliamentary Opposition, claiming that the public relations campaign from the mining lobby had emboldened climate change sceptics with the Coalition to oppose their leader and his support for the Rudd government's ETS (Grigg 2010a, b, c).

With Tony Abbott in charge of the parliamentary Opposition, the Rudd government's final attempt to pass the CPRS was defeated in December 2009, just before the unsuccessful UN conference on climate change in Copenhagen. By April 2010 the Rudd government publicly dropped any further proposals for an emissions trading scheme. This back down on climate policy reform damaged Prime Minister Rudd's credibility with large parts of the Australian electorate. His popularity and leadership were further damaged after the mining lobby launched a new advocacy advertising campaign against the government's proposal for a profits tax for the mining industry (McKnight and Hobbs 2013). This second campaign by the mining lobby further eroded the prime minister's approval rating in the opinion polls and precipitated a leadership crisis in the government. Events came to a head in June 2010, with Kevin Rudd replaced as leader (and Prime Minister) by his deputy, Julia Gillard (McKnight and Hobbs 2013).

Following Rudd's removal, Prime Minister Julia Gillard proceeded to negotiate a much more modest mining tax with the mining lobby in order to end their advertising campaign against the government. The design of this new tax strongly favoured the commercial interests of the big multinational mining conglomerates that had been funding the industry's advertising campaign, by including different forms of tax concessions for "mature projects" (Cleary 2011, p. 78). Having negotiated a truce with the powerful mining lobby, Gillard then moved to consolidate her leadership as prime minister by calling a federal election for August 2010.

However, Gillard's truce with the big multinational mining conglomerates would not last. A close election resulted in a 'hung parliament', an outcome which gave neither of the major political parties enough seats in the House of Representatives to form a government in their own right. The inconclusive result in the election, forced Gillard to build alliances with independent parliamentarians and the Greens party in order to return

Labor to government. Yet the price of Greens support was a fixed price on carbon, or 'carbon tax'. This second attempt to price carbon would also meet fierce resistance from the fossil fuel lobby in the form of a third persuasive PR campaign when it was introduced in 2011, and is discussed in Case Study 3.

## CASE STUDY #2: THE CHARM OFFENSIVE, 2010–2011

The mining lobby's public relations campaign against the Rudd government demonstrated the communicative power of a well-funded and executed public relations campaign. At the time, advertising and PR executives praised the campaign for its targeted messaging, which 'successfully sold' the anti-tax message by appealing to 'empathy', 'ideas of fairness' and 'solidarity' with an industry that had 'saved the nation' from the global financial crisis (McKnight and Hobbs 2013). While this campaign was led by the Minerals Council of Australia (MCA) and funded by the big multinational mining conglomerates, its strategic success was largely the result of the creative talents of Neil Lawrence and his advertising agency, Lawrence Creative—which is somewhat ironic given that Lawrence had been the 'creative architect' behind Kevin Rudd's successful 'Kevin 07' election campaign. Lawrence's work with the MCA continued after Rudd's political demise, with his firm producing a series of advocacy advertisements that clearly sought to bolster the industry's public image as part of a proactive inoculation strategy.

It is possible that many Australian public relations executives have not heard of inoculation public relations, and if they have it might seem too 'academic' for Australian practitioners that often come from careers in journalism, marketing or politics rather than having studied public relations at university (Harrison 2011). While the term 'inoculation PR' is not in common use, the strategy it signifies is widely used in Australia, but is colloquially referred to as a 'charm offensive'. As the name suggests, a charm offensive seeks to build relational capital with a range of publics, which then become allies or at least ambivalent or neutral about the organization's operations. Much like the inoculation strategy in the US, the charm offensive consists of community sponsorship programs, political donations and advocacy campaigns, which seek to publicise the socioeconomic benefits of a specific corporation or industry. As Richard Denniss (cited in Cleary 2015, p. 30) notes, in Australia the mining industry has became very adept at taking full credit for the indirect benefits of mining while distancing themselves from any responsibility for the indirect costs.

Following their aggressive and successful campaign against the Rudd government in 2010, the mining lobby decided to increase its proactive charm strategies of community sponsorship and engagement. Neil Lawrence was once again employed by the Minerals Council to create a mass advocacy advertising campaign, with the goal of reframing some of the negative issues associated with the industry. To this end, Lawrence Creative refreshed and enhanced the MCA's campaign 'NewGenCoal', creating a website and video to promote the science of carbon capture and storage (CCS), or 'clean coal', as the technological solution to climate change. The website featured scientists and industry experts who provided upbeat interviews about the powers of CCS technology. However, their statements failed to disclose that many of these 'independent' experts were in fact funded or employed by the coal industry (Pearse et al. 2013, p. 103).

This coal-specific campaign was greatly enhanced by a new national campaign that began in early 2011. Under the banner 'Mining: This is Our Story', Lawrence Creative produced a series of sophisticated advertisements for broadcast, print and online media, which convincingly positioned mining as a central support for the Australian way of life. These advertisements focused on the individual and personalized stories of 'everyday hardworking Australians', and sought to actively counter the perception that mining was an industry dominated by white, working class males. One of these lengthy advertisements focused on champion cyclist Anna Mears, and how BHP Billton's sponsorship and support had inspired her comeback from a bad cycling accident. Another featured the story of 'average mining employee' Heather, whose employment with the mining company allowed her to live the Australian dream of career, family, tropical lifestyle and harmonious home life. Other advertisements promoted the mining industry's positive impact on the lives of migrant workers, featuring the stories of men and women from Asian, Islamic and French backgrounds. Another advertisement sought to counter public perceptions regarding potential conflicts between mining companies and Indigenous Australians, by featuring the story of an Indigenous chemical engineer whose career in the mining industry was allowing him to live a life that made his mother 'proud'.

Collectively these advertisements, which ran from early 2011 to mid-2012, were powerful communicative tactics that effectively built the myth that the mining industry was central to the Australian economy and way of life. They were also widely seen by the public. In a corporate video highlighting the work of his agency, Neil Lawrence stated that nearly 17 million Australians (out of a national population of 22.5 million) had seen the televised versions of his advertisements, on average eight times, with a further 8.9 million also being exposed to the print media versions (Pearse et al. 2013, p. 102). Many others saw the advertisement in cinemas or online. According to Lawrence's market research on the campaign's effectiveness, these advertisements made people 'feel good about mining' and 'want to get a job in mining'. Moreover, much like his early campaigns 'Kevin 07' and 'Keep Mining Strong', fellow-advertising executives praised the 'Mining: Our Story' advertisements for their creativity, visuals and emotional impacts. However, this would be Neil Lawrence's last campaign for the Minerals Council of Australia, as he was killed in a scuba diving accident while on holiday in the Maldives in 2015 (Zwartz 2016).

The charm offensive of the mining industry does of course consist of other tactics than advocacy advertisements. As Pearse et al. (2013) have documented, the mining industry has also been employing community-targeted public relations in the form on sponsorships, grants, education resources and other financial largesse, which are often celebrated in Corporate Social Involvement brochures. Some of these activities include aid relief for families affected by floods, funding helicopter rescue services and healthcare facilities at regional hospitals, sponsorships for cultural programs such as wine and food fairs, film festivals and agricultural shows, producing education resources on the mining industry for schools, providing money for conservation programs for threatened species and environmental restoration and community support grants for services offering affordable housing. In addition to these benevolent efforts, which further bolster community support for coal mining, industry bodies also provide financial support for researchers and think tanks such as the Melbourne-based Institute of Public Affairs (IPA), which has been supporting the mining sector with research and reports that promote doubt about climate change (Pearse et al. 2013, p. 150).

The final aspect of mining industry's charm offensive in recent years has been their financial donations to individual politicians and political parties (Murphy 2016; Slezak 2016). While contradictory disclosure laws and loopholes make it difficult to identify all political donations provided by the mining industry to Australia's democratic process (Orr 2007), federal donations are disclosed to the Australian Electoral Commission and show that the mining and energy sectors have increased their political donations to federal politicians in recent years (Ting and Begley 2015).

As can be seen in Fig. 10.1, contributions from the mining and energy companies have favoured the two major conservative parties, the Liberals and Nationals, which together form the parliamentary alliance, the Coalition. The sums in Fig. 10.1 do not, however, include all political donations made by the mining sector. Organizations can donate funds to party affiliated organizations, which can make it difficult to easily identify how much money has been donated by an entire industry to the respective political parties. For instance, donations by two of the biggest mining conglomerates, BHP Billiton and Rio Tinto, are not included in Fig. 10.1 and yet in the 2014–2015 financial year these corporations collectively gave over $200,000 to the Liberal party-linked Cormack Foundation (Holmes 2016). While not an exhaustive list, Fig. 10.1 captures most of the federal donations provided by the mining and energy companies operating in Australia and provides broader insights into the direction of their political donations in recent years, with the figures showing the industry's

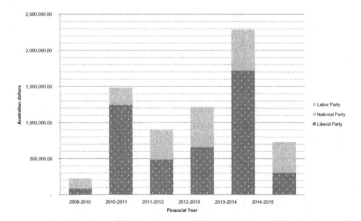

**Fig. 10.1** Federal political donations made by mining and energy sector. *Source* Compiled from Australian Electoral Commission reports. *Acknowledgment* Eugenia Lee, Research Assistant, University of Sydney. For a breakdown of the individual companies included on this list see Ting and Begley (2015).

opposition to the Labor government's policies to reduce carbon emissions. It is to the final attempt by the Labor government to implement a carbon price that we now turn.

## CASE STUDY #3: INTRODUCTION OF A CARBON TAX, 2011–2013

The third case study concerns the campaign mounted by the coal industry, along with large sections of business, to oppose a carbon tax during the period of the Labor government of Prime Minister Julia Gillard (2010–2013). The carbon tax emerged as the price of the aforementioned alliance between the minority Labor Party and a group of independent members of parliament, including a Greens MP. As a consequence of the horse-trading involved, proposals for a carbon tax (actually a fixed price for carbon emissions for 3 years after which the price would 'float' as part of an emission trading scheme) came with a number of other measures including cheap credit and other support for renewable energy projects, along with an independent watchdog on targets for carbon emissions.

While the government was debating the details of these energy policies during most of 2011, the coal industry and its allies publicly condemned any carbon tax, arguing that it would mean the loss of anything between 10,000 and 23,000 jobs and the closure of 16 coal mines (Crowe 2011). In June it began an advertising campaign with the message "Why should thousands of Australians lose their jobs if it makes no difference to global greenhouse emissions?" (Australian Coal Association 2011). As with the campaign against the previous government's emissions trading scheme, the advertisements targeted regional mining areas and their members of parliament to create a sense of looming unemployment. The industry was supported by the Opposition leader, Tony Abbott, who told a conference of mining companies, "you need to become political activists" to fight the carbon tax (Coorey 2011). The popular media were seen as a key battleground for public hearts and minds. In a letter to members, the chief of the mining industry lobby group said that "the new paradigm [of policy development] is one of public contest through the popular media more so than rational, considered, effective consultation and debate" (Priest 2011).

The result was the formation in July 2011 of a new and broad business group, the Australian Trade and Industry Alliance (ATIA), which was composed of mining, retail and manufacturing bodies. With such broad based support, the mining and coal industries would be seen as less sectional. The ATIA then jointly launched a $10 million dollar advertising campaign opposing the carbon tax. The leaked casting brief for the advertisements argued that the characters featured must look like 'people who represent Australian families, small business owners, Australian workers and retirees' (Priest 2011). The advertisements appeared under the heading "Carbon Tax Pain, No Climate Gain". They anchored in personal stories from people like 'Angela, a hairdressing salon manager', who was quoted as saying it was 'the worst possible time to introduce a carbon tax' (Australian Trade and Industry Alliance 2011). The tax itself was described as "the world's biggest carbon tax". Among the claims made in the advertisements was that the cost of electricity would rise 20%, a claim later found to be misleading and having no 'independent verification' by the national competition and consumer regulator (Priest and Walsh 2012).

The Clean Energy Future package of laws was passed by parliament at the end of 2011 and involved introducing a $23 per tonne carbon tax to be paid for by the top 500 polluting corporations starting in July 2012. The head of the mining lobby, Mitchell Hooke, said parliament has voted "to reduce Australia's standard of living" to reduce global competitiveness and to cut jobs (Maher 2011). The doom-laden predictions were also undermined by the fact that, as the carbon tax was introduced, investment in coal mining boomed. As one report noted: "the impending carbon price has done nothing to deter investment in the coal industry with spending on exploration surging faster than any other mineral commodity" (Ker and Morton 2012). By this time, however, the conservative Opposition leader had enthusiastically taken up opposition to the tax. Tony Abbott made the abolition of the 'toxic tax' one of the central issues in his campaign to defeat the government. This succeeded at the national election in September 2013 and his government abolished the carbon tax shortly afterwards.

## The Struggle Continues

Following their victories over the ETS and the carbon tax, the coal industry has continued to produce sophisticated PR campaigns intended to protect their operations. In 2015, the industry launched the advertising and PR campaign called "Little Black Rock", which promoted the phrase 'coal is amazing'. This campaign, developed by the MCA, emphasised the economic value of coal exports and claimed that new types of coal burning power stations had low carbon emissions (Evans 2015). It also marked the beginning of new inoculation PR campaign in that it sought to bolster the industry image in the lead up to the climate change conference in Paris in late 2015.

However, the advertisements were criticised for their over-the-top claims, with the campaign widely perceived to have failed after polling showed a drop in approval for coal (Meade 2015). The president of the coal division of the Anglo-Australian mining and petroleum conglomerate, BHP Billiton, Australia's largest company, acknowledged that the coal industry was "losing the public relations battle against activists who are trying to shut down fossil fuels" (Ludlow 2015) and called on the global mining industry to "step up and help improve the quality of debate". At the same time, one of the world's biggest PR firms announced that it would no longer work with coal producers and climate change deniers (Goldenberg 2015).

## CONCLUSIONS

Inoculation and persuasive PR strategies are often employed by industries that can suffer from crisis of declining social legitimacy due to changing social and political values and concerns. In Australia, these strategies have been used by the mining industry to defeat governments seeking to reduce carbon emissions. Political donations and internal lobbying have been used to build relationships with influential politicians. Where parliamentarians have not been receptive to industry concerns, the lobby groups and their allies have taken their arguments to the media with the intent of bring pressure to bear on individual politicians who must face re-election from agitated constituents.

The mining industry has used this reward and punishment approach to great effect, spending tens of millions of dollars on advocacy advertisements and other PR tactics in order to bolster their public image and attack climate change policy proposals with which they disagree. Such persuasive strategies are not necessarily subtle or polite. Reflecting on the campaign against the carbon tax, the former Labor Treasurer, Wayne Swan, said the government had come up against powerful vested interests which were determined to "smash us and smash us and smash us again ... when you've seen the way they operate behind the scenes, through climate change and through mining tax issues, they are brutal, they are powerful, they are selfish, they take no prisoners" (Chubb 2014, p. 188).

In response, PR practitioners would likely argue that their role is merely that of an advocate in the court of public opinion, pleading the case for the 'other side of the debate' in order to counterbalance claims from those working for environmental lobby groups or academics in research institutions. But this courtroom analogy ignores the significant communication power of the mining companies. The public sphere is not structured like a courtroom and there is no magistrate to ensure a fair hearing and rational deliberation. Moreover, unlike a public health and safety campaign, the usual justification for persuasive PR campaigns—that the ends justify the means—does not apply here. The scientific consensus on climate change shows that the world needs to transition to greener forms of energy production and soon. Communication campaigns that stymie this transition endanger us all, including the employees working and living in mining regions.

# CHAPTER 11

# Bearing Witness and the Logic of Celebrity in the Struggle Over Canada's Oil/Tar Sands

*Patrick McCurdy*

## INTRODUCTION

Although Canada's bitumen sands have been extracted on an industrial scale since the 1970s, they did not enter the public consciousness as a contentious issue until decades later. The current wave of protests against the tar sands may be traced back to 2005 and campaigns run by environmental non-governmental organisations (eNGOs) such as Greenpeace, World Wildlife Fund (WWF) and Forest Ethics (Katz-Rosene 2017). However, if you had to pick a date as to when contention over the tar sands really entered the public imagination, it would perhaps be January 17, 2008. On that day, a group of activists from Stop the Tar Sands erected a sculpture in front of Calgary's Telus Conference Centre where the fifth annual Canadian Oil Sands Summit was being held. The installation depicted three Alberta government politicians—represented by three stuffed men's suits—each with their head buried deep in a large pile of symbolic 'tar sand'. A briefcase lay in place of the submerged head of each scarecrow-like figure each clearly labelled with a politician's name:

P. McCurdy (✉)
Department of Communication, University of Ottawa, Ottawa, Canada
e-mail: pmccurdy@uOttawa.ca

© The Author(s) 2017                                                131
B. Brevini and G. Murdock (eds.), *Carbon Capitalism and Communication*,
Palgrave Studies in Media and Environmental Communication,
DOI 10.1007/978-3-319-57876-7_11

"Stelmach" for Alberta's Conservative premier, "Knight" for his Energy Minister and "Renner" for his Environment Minister. A long black sign was placed on the ground in front of the installation; white capital letters on the sign read "Get your head out of the tar sands".

This symbolic protest was held at the cusp of what has become a persistent and international campaign against bitumen development and helped establish the 'tar sands' frame. In the time since these protests, activists and other civil society stakeholders have manufactured and exploited numerous political opportunities to contest tar/oil sands development in the name of environmental protection, Aboriginal rights, and stopping climate change. Meanwhile, industry stakeholders along with the Alberta provincial and Canadian federal government have launched their own campaigns, initiatives and counter narratives (Turner 2012).

This chapter views struggles over bitumen development and the struggle over climate change as simultaneously material and mediated. From this perspective, media—both social and legacy—function as key arenas, resources and sites of information in the ongoing battle for the public's imagination (Castells 2009). In exploring this struggle, I want to apply and extend Cammaerts' (2012) concept of the mediation opportunity structure to examine two broad activist logics which underwrite bitumen sands campaigning. The first, the 'logic of bearing witness', operates through the creation of 'image events' (Deluca 1999) employing spectacle and symbolic direct action at oil/tar sands plants, mines and public events. Second, 'the logic of celebrity' examines how celebrities' visits to the Alberta's bitumen sands and their remarks about them relate, inform, underwrite and impact 'tar sands' activism using case studies of Neve Campbell, James Cameron, Neil Young and Leonardo DiCaprio.

## MEDIATION AND POLITICAL OPPORTUNITY

Contemporary political struggles are inevitably also mediated struggles in which media representations of social and political issues operate as both sites and sources of political contention (Castells 2009; Couldry 2013; Silverstone 2007). Social movement scholars use the concept of 'political opportunity' to identify moments of contention and their mobilization by social movements (Tarrow 1998). However, as Cammaerts (2012) has rightly argued, the role of media and communication within this perspective is woefully undertheorized. Addressing this lacuna, he has proposed the idea of 'mediation opportunity structures' to conceptualize the role of media in relation to protest and social movements. Of particular interest is his focus on the media-savvy repertoires of contention enacted by political actors.

In developing his perspective Cammaerts cites della Porta and Diani's (2006, p. 17) argument that there are "three distinct, but not mutually exclusive, logics that activists ascribe to their protest actions when deciding from the available repertoire of action" (Cammaerts 2012, p. 121). Each of these logics is underwritten by a reflexive awareness, or at least a 'lay theory', of which actions will receive media coverage and what kind of coverage it is likely to be (McCurdy 2011).

First, the "*logic of numbers*", is most commonly expressed through mass demonstrations. As a tactic for claiming attention however, filling the streets with marchers has become routinised and predictable in both its execution and the media reporting it attracts.

The second logic, the "*logic of damage*", involves the threat or enactment of damage to either property and or/people. Here, Cammaerts (2012) rightly stresses the need to clearly differentiate both "conceptually and morally" between "violence—directed at people—and damage—directed at property" a differentiation too often lost in the resort to the catch-all label, 'political violence' in mainstream media discourse.

In contrast, the third logic, the "*logic of bearing witness to injustice*" is firmly rooted in principles of non-violence. As della Porta and Diani argue, protest action underwritten by this logic is "…not designed to convince the public or decision makers that the protestors constitute a majority or a threat. Rather it seeks to demonstrate a strong commitment to an objective deemed vital for humanity's future" (2006, p. 176). As exemplified by the head-in-the-sand installation described earlier, acts of witness now routinely incorporate resonant images or spectacular and/or performative elements. Their currency is evident in the rise of media stunts and "image event" (Deluca 1999) protests. However, I would argue there is a fourth logic not initially discussed by della Porta and Diani (2006) nor expanded upon by Cammaerts: the *logic of celebrity*.

The logic of celebrity is a modus operandi symptomatic and emblematic of living in media-saturated times. Recognizing celebrity as a logic acknowledges the extent to which, under contemporary conditions, the use of, or resistance to, celebrity involvement in protest can underwrite, orient, constrain or facilitate political action. The logic of celebrity operates whenever established celebrities, such as actors and musicians, are recruited to boost the visibility of a social cause and in the cultivation of "celebrity activists" (McCurdy 2013) such as Julian Assange or Edward Snowden. In a media system such as ours, which manufactures, perpetuates, trades in, sells and feeds off celebrity, celebrities are given a privileged platform that guarantees that even their passing comments or off the cuff remarks are given elevated attention, status and scrutiny.

The use of celebrity to advance both the logic of numbers and logic of bearing witness is well established in movement repertoires of political contention, becoming a staple in activist campaigning and environmental and climate movements in particular (see Brockington 2009; Meister 2015). Celebrities can act as both a resource—a form of capital—and a catalyst in mobilizing political actors as part of the logic of numbers by encouraging the public to attend a mass demonstration and by endorsing or promoting the event in advance and/or participating themselves. Their presence can also amplify the visibility of actions which seek to bear witness to injustice by heightening the 'newsworthiness' of the event and thus the likelihood of media coverage. At the same time, employing celebrity may also impede other actions, creating tension within movements over focus, objectives and power (Gitlin 1980; Meyer and Gamson 1995). However given the fact that celebrities—rightly or wrongly and for good or for ill— are afforded political standing in their own right, we need now to recognise celebrity as an additional and separate logic that may hinder as well as advance the causes and claims they are enlisted to support. Consequently, this chapter examines the interplay between the logics of celebrity and bearing witness through a case study of the mediated battle over Alberta's bitumen.

The empirical material referenced and drawn upon here is taken from the larger Mediatoil research project conducted by the author (see www.mediatoil.ca).[1] Mediatoil documented the evolution of contention over the oil/tar sands as captured in stakeholder's promotional material. Part of the documentation process involved creating a timeline of events relating to the bitumen sands. This produced 179 unique entries documenting project and government announcements, significant meetings and milestones, notable publications and instances of protest. The analysis presented here focuses on the 96 entries related to the political contestation of the Alberta's bitumen sands between May 2007 and January 2016. While major efforts were made to document as many relevant bitumen-related events as possible the Mediatoil timeline may not be exhaustive. It is however, comprehensive enough to allow us to chart the development of political contention around the oil/tar sands over the course of almost a decade.

## BEHIND THE BITUMEN BATTLE: AN OIL/TAR SANDS PRIMER

The Canadian bitumen sands, also referred to as oil sands or tar sands, are a vast (142,200 square kilometre) deposit of sand, clay sand and a type of heavy oil known as bitumen. The tar sands are the third largest proven resource of oil in the world at 166 billion barrels of oil (Government of Alberta 2016). Currently, the oil sands produce 2.4 million barrels of oil per day (BPD), which accounts for 55% of crude production in Canada (Government of Alberta 2016; CAPP 2015a). Bitumen production is anticipated to grow, with the Canadian Association of Petroleum Producers projecting 3.08 million bpd by 2020 and 4 million bpd by 2030 (CAPP 2017, p. ii).

Unlike conventional light crude oil, bitumen is incredibly resource intensive to refine and requires a massive and extremely costly supporting infrastructure. There are two main methods for extracting bitumen: open-pit mining and 'in situ' (Latin for in place). Open pit mining is like coal mining, where bitumen is extracted from the deposits up to 75 meters underground using massive machinery. Currently such mining projects account for 1.1 million BPD in Alberta. However only 20% of Alberta's oil sands reserves are accessible by this method (CAPP 2015a). The other method for extracting bitumen is referred to as 'in situ' and is used for bitumen which cannot be reached by conventional mining. Alberta currently produces 1.3 million BPD using in situ methods—the most common of which is Steam Assisted Gravity Drainage (SAGD), and 80% of Alberta's bitumen reserves must be accessed by this method (CAPP 2015a).

Just as refining bitumen is capital intensive, it is also resource intensive. The well-to-wheels emission intensity of oil sands crude is between 8 and 37% higher than conventional crude depending on the extraction method (Pembina 2012). The increased greenhouse gas (GHG) intensity of oil sands production together with the volume of both natural gas and fresh water used to produce oil from bitumen and the impact of production on the surrounding environment and communities have been significant issues of contention for oil sands opponents (see Council of Canadians 2015; Greenpeace 2015). Responding to concerns, the industry claims that "Oil sands GHG emissions have declined 30% per barrel from 1990 to 2013" (CAPP 2015a). There have been per barrel efficiencies, but as the environmental NGO (eNGO) Pembina asserts, "Oil sands are the fastest-growing source of greenhouse gas emissions in Canada" (Dyer et al. 2013). Moreover, oil sands GHG emissions are projected to increase by 124% from 2010 to 2030, rising from 64 megatons to 115 megatons (Fekete 2016).

With the planet on the precipice of climate calamity, environmental and climate change campaigners have zoomed in on the projected increases in oil sands production and emission as a political, symbolic and discursive opportunity to discuss climate change and tie the issue of oil sands production to our unsustainable carbon-based lifestyles. Early campaigning against Alberta's bitumen sands involved "image events" (Deluca 1999) such as banner drops at open pit bitumen mines and symbolic protests in front of or even inside oil-related political events and conferences.

## IMAGE EVENTS AND MEDIATED WITNESS

In his book *Image Politics*, Kevin Deluca looks at the rise of "image events" as a form of political expression used by environmental organizations. These interventions, which operate as "staged protests designed for media dissemination" (Delicath and Deluca 2003, p. 1), have been studied in the context of, among others, anti-globalization protests (McCurdy 2011) and climate change campaigning (Greenberg et al. 2011; Askanius and Uldam 2011; Doyle 2011).

Many early protests against the oil sands took the form of image events. Fittingly, it was Greenpeace Canada who seemed to be the most active in early tar sands protests. Fitting because it was Canadian Bob Hunter who founded the edgy eNGO and who is credited with inventing the idea of "media mind bombs" (Hunter 2004; Zelko 2013); protest actions designed to capture the media's attention and public's imagination.

Whether referred to as mind bombs or image events, this tactic is underwritten by a logic of bearing witness but with a media mindfulness. An early example from tar sands contention is Greenpeace's unfurling of a banner on April 23, 2008 during Alberta Premier Ed Stelmach's annual Premier's Dinner at the Shaw Center in Edmonton, Alberta. A Greenpeace video of the incident posted on YouTube shows an activist suspended from the conference center rafters unveiling a black fabric banner. In frame, alongside the activist is a giant television screen showing Premier Stelmach thanking the crowd for their applause (Greenpeace 2008b). Eventually the attention of both media and security in the Shaw Center turn to the activists and their banner which read in big white font "Stelmach: the best premier oil money can buy" (Greenpeace 2008b). The video then shows security swiftly escorting Greenpeace protesters from the premises encircled by a roaming scrum of 8–10 journalists, photographers and cameramen. While security dealt with the interruption to the Premier's fundraising dinner promptly and professionally, the stunt received national media attention, successfully disrupting the event's intended framing and creating a mediated opportunity for oil sands contention (cf. Le Billion and Carter 2012).

On July 23, 2008—exactly three months after the Stelmach protest—Greenpeace activists turned their attention to Syncrude's Aurora North operation. The site was targeted as 1600 ducks had previously died at that location after landing in a tailings pond.[2] Pictures of the "dead ducks" would become powerful mind bombs in activists' repertoire of images underlining the toxicity of the tar sands. The July banner drop at the site was executed with semiotic precision. Activists sought to "transform the opening of a tailings pond pipe into the 'mouth' of a giant skull spewing toxic sludge" (Greenpeace Canada 2008a). A second, significantly bigger white banner with large black font was positioned against an upper bank of the tailing pond read: "World's dirtiest oil; Stop the tar sands" (Ibid.). Both banners were simple but strong in their messaging. The visual and performative nature of these direct actions integrates an acute appreciation of mediated communication into the logic of bearing witness to injustice, amplifying the original message with an online press release featuring photographs of the event and videos of the action uploaded to YouTube (see Greenpeace 2008a).

In the eight plus years since this intervention, Greenpeace and other eNGOs protesting the tar sands have engaged in a range of further direct actions. In August 2011 Tar Sands Action drew on the logic of numbers and logic of bearing witness to undertake a mass sit-in in Washington, D.C to oppose the Keystone XL pipeline. The protest, which ran from August 20 to September 3, 2011, was designed and executed as a symbolic and coordinated act of civil disobedience (McKibben 2011; Wihbey 2011). The event's location—the gated and heavily policed sidewalk of 1600 Pennsylvania Avenue—ensured that a clear and centered view of the White House served as a backdrop for media documentation of the sit-in and the subsequent arrest of participants. In addition, some activists reinforced their message with visual props and homemade and professional banners with slogans such as "Lead on Climate", "No tar sands XL pipeline",

"We > Tar Sands" and "Offshore wind not tar sands oil". On the first day of the action 70 protesters were arrested for "failing to obey a lawful order" while a total of 1,252 arrests were made over the course of the two-week sit-in (Polden 2011; Tar Sands Action 2011; Tackett 2011).[3]

On the heels of the Washington sit-ins, a group of Canadian eNGOs including The Council of Canadians, Indigenous Environmental Network and Greenpeace Canada issued a call for a similar sit-it style protest action to be held on September 26, 2011 to oppose the Keystone XL pipeline (Adeland 2011; Kraus 2011). The one day event was promoted as potentially one of the "largest acts of civil disobedience on the climate issue that Canada has ever seen" (Kraus 2011). Like its Washington counterpart, the location for the Ottawa sit-in, Parliament Hill, Ottawa, was imbued with symbolic power and political significance.

While the initial plan was to demonstrate near the House of Commons, anticipating this move, the Royal Canadian Mounted Police (RCMP) pre-emptively constructed security fences on Parliament Hill in front of the iconic Center Block limiting building access. These metal barricades marked the territorial divide between 'legal' and 'illegal' protest.[4] Determined to engage in civil disobedience, activists approached and scaled the waist high 'perimeter fence' in waves. The first line of protestors advanced to the chants and drums of Aboriginal supporters in attendance together with the cheers and applause of other attendees. The actions of those scaling the symbolic fence were followed and documented by a roving pack of at least 30 journalists, cameramen and photographers pressing against and jostling with each other to document the event. In total, the RCMP made 117 arrests. The Council of Canadians, a primary event organizer assessed its success in terms of media coverage asking, in a blog post, "How well did mainstream media cover our tar sands protest?". Their response: "The protest on Parliament Hill yesterday was a tremendous success. It was also extensively covered (in a mixed way) by Canada's mainstream media" (Council of Canadians 2011).

Ottawa's "success" and, indeed, Washington's, was rooted in two overlapping factors. First, activists were fully committed to bearing physical witness to their goal of stopping bitumen pipelines and prepared to get arrested for this objective. Second, they were attuned to media logics and conducted their actions in full media view successfully creating, capitalising on, and in the case of Ottawa, extending, a mediated opportunity.

Both the Washington and Ottawa sit-ins also mobilized celebrity involvement and endorsement. In Washington D.C., actor Darryl Hannah and climate change campaigner Bill McKibben were both arrested. In Ottawa, media coverage prior to the event heralded the presence of actors and activists together with headlines such as "Protesters? Check! Police? Check! Celebs? Also check!" (Smith 2011). In both instances the presence of celebrities amplified the mediated logic of bearing witness by increasing the event's newsworthiness. The next section explores the logic of celebrity in oil sands campaigning in greater detail.

## CELEBRITY CAPITAL AND TAR SANDS PROTESTS

Protests around Alberta's bitumen sands have consistently attracted celebrity visits, interventions and acts of bearing witness (Wilt 2015). Of the 96 tar sands protest actions coded for this research celebrities were involved in 39 (40.5%) of the political contention undertaken.

The honour of the first celebrity visit to Alberta's tar sands goes to Canadian actress Neve Campbell. Campbell, a former A-list Hollywood star, was known for her appearance in the popular 1990s television show *Party of Five* as well as the horror movie *Scream* (1997). In October 2008 as part of an 'awareness raising strategy' eNGO ForestEthics (rebranded Stand in April 2016), organized a two-day tour where Ms. Campbell visited Fort McMurray, various bitumen sites and met with the Chiefs of the Mikisew Cree First Nation [MCFN] and the Athabasca Chipewyan First Nation (ACFN) (Nease 2008). Campbell's comments that she was "horrified by the pace and scale of development in the tar sands, and the weak response by our federal and provincial government" made national news (ibid.). Campbell's staged visit can be seen as the first of many celebrity pilgrimages to bear witness to the oil sands designed to increase exposure, reach and awareness of the issue. Although there is a degree of both play and irony in the horror movie actress' remarks about being "horrified" at the oil sands.

In late September 2010 Canadian James Cameron, an Oscar-winning film director whose credits include *Terminator, Aliens, Titanic* and *Avatar* visited Alberta's bitumen sands. His three-day excursion was initiated after an April 2010 invitation from George Poitras on behalf of the Indigenous Environmental Network (Tarbotton 2010). The invitation was made during the United Nations Permanent Forum on Indigenous Rights and spurred by Cameron's comments that the oil sands were a "black eye" to Canada's image as an environmental leader (Canadian Press 2010). Four months prior to the forum, Cameron was in the news for the Christmas

2009 release of his science fiction film *Avatar*. The high-budget Hollywood blockbuster was a feature length environmental parable which pitted space travelling money-hungry humans keen to extract resources across the universe against the Navi, the indigenous peoples of Pandora, determined to protect their planet from being plundered by the interstellar extractive industries. Immediately upon the film's release parallels were drawn between the struggle over Pandora's resources and over Alberta's tar sands (see CBC National 2010; Haluza-DeLay et al. 2013; Itzkoff 2009). *Avatar* was further linked with tar sands campaigning when a group of over 50 eNGOs spent $20,0000 USD on a full page ad in *Variety* magazine's March 2010 Oscar issue endorsing *Avatar's* nine Academy Award nominations with the headline "JAMES CAMERON & AVATAR YOU HAVE OUR VOTE. CANADA'S AVATAR SANDS" (Barnes 2010; Rowell 2010, emphasis in original). Consequently, when Cameron finally visited Alberta in September 2010, he did so in the wake and context created by his film. Cameron's trip included a meeting with Alberta Premier Ed Stelmach, a tour of Syncrude and meetings with First Nations leaders (Wingrove 2010). The media space created around Cameron's tour was described by Elizabeth May, leader of Canada's Green Party, as follows:

> What James Cameron has done has been to help a lot of Canadians take the tour with him, to see the massive scale, to see the devastation of the environment, to understand more of the science and to see where we really are right now on this planet in terms of our overall imperative to get off fossil fuels (CTV News Staff 2010).

Cameron's choreographed celebrity pilgrimage to the tar sands capitalized and extended the mediated opportunity created by *Avatar*. By travelling to Alberta to bear witness to the tar sands, Cameron opened mediated opportunities for activists and First Nations representatives to secure a public platform for debate and contest. His Hollywood celebrity status, the relevance to the issue of his hugely successful film, *Avatar*, and his status as a Canadian citizen combined to provide a powerful news hook to raise awareness of the tar sands into the public imagination.

Canadian musician Neil Young has been one of the most vocal celebrities to speak out against the tar sands and has payed particular attention to the land rights issues of First Nations. Young's political activism and polemic statements have garnered national headlines and made him into a lightning rod for critics. On September 9, 2013, while attending a National Farmers Union event in Washington, D.C., he commented that, "The fact is, Fort McMurray looks like Hiroshima. Fort McMurray is a wasteland" (Koring and Cryderman 2013). His remarks made news across Canada and spurred a wave of defensive and angry responses from oil sands supporters and local politicians. The mainstream media editorials that followed framed Young as a hypocrite who indulges in an oil-soaked celebrity lifestyle (e.g. Lamphier 2013). A local radio station in Fort McMurray even declared a "No Neil" day banning Young's music for 24 hours in protest (Canada Press 2011).

In January 2014, Young used his celebrity status to organize a four city concert tour—*Honour the Treaties*—to raise awareness and funds for ACFN in their struggle against tar sands development. The concerts generated revenues of $500,000 but also incurred the ire of critics (Ball Windspeaker 2014). Oil sands advocate Ezra Levant and his organization *Ethical Oil* started an online website and campaign "Neil Young Lies" which sought to "[expose] Neil Young's 'honour the treaties' hypocrisy, and how the aging rockstar really feels about First Nations" (Neil Young Lies 2016). The website sought to 'fact check' Young's oil sands claims, hypothesize who funded Young's concert tour and, expose "the truth about Neil Young" as seen in the hypocrisy of Young's pitching his humanist and ecological sensibilities against his celebrity lifestyle (ibid.). Focussing on a celebrity's use of oil-based products is a common strategy of ad hominem attack when celebrities speak about environmental issues. Reflecting on the controversy stirred up by his comments and tour Young remarked, "My job is to bring to light the situation through my celebrity" (CBC News 2014). In this quote Young recognizes his celebrity as a form of capital which he can mobilize and direct to generate media attention and economic capital.

## SLIPPAGE: GLOBAL CLAIMS AND LOCAL KNOWLEDGE

The only other major celebrity who has been as visible as Young in campaigning against the oil sands is Hollywood actor and climate change campaigner Leonardo DiCaprio. Whereas both Cameron and Young were born in Canada and could plausibly claim to have a stake in the future of the country and environment they grew up in, DiCaprio's status as an American citizen opened a potential space for arguments that he was disconnected from and ignorant of local conditions. As we will see, an ill-advised remark given in interview allowed supporters of tar sands development to bring the full force of this argument into play. His initial intervention however, attracted more positive publicity.

In August 2014, he toured Fort McMurray in the course of making the National Geographic channel documentary *Before the Flood*. While the visits of Campbell, Cameron and Young involved some form of press conference, DiCaprio's visit did not. However, some coverage was posted to social media by tar sands campaigners and then picked up by local and national media (McDermott 2014). While touring Alberta, DiCaprio accepted an 'Ice-bucket Challenge' from British football star David Beckham. The Challenge was part of a viral media campaign where individuals were filmed having a bucket of ice water dumped over them to raise awareness about amyotrophic lateral sclerosis (ALS) and raise funds for research. Once completed, the individual could then 'challenge' three others to complete the same task, share the video and his/her subsequent nominations. The campaign gained particular traction through the involvement of high profile celebrities.

DiCaprio's Ice Bucket video was filmed on the shores of Lake Athabasca, Alberta, in the company of about 50 members from the ACFN, MFCN and tar sands campaigners (Isis1174 2014). While DiCaprio participated in the ALCS campaign, he simultaneously used his video as a political platform to amplify tar sands issues. He achieved this by incorporating challenges from Chief Adam of the Athabasca Chipewyan First Nation who challenged CAPP President Dave Collyer, Chief Courtoreille of the Mikisew Cree First Nation who then challenged Suncor Executive Vice-President Mark Little, and Sierra Club President Michael Brune who challenged Shell's CEO Ben van Beurden. Meanwhile, DiCaprio challenged Canadian Prime Minister Stephen Harper (Isis1174 2014). DiCaprio's politically calculated Challenge video received national news attention. Although Prime Minster Harper did not reply directly, DiCaprio was able to use his celebrity status to draw attention to the tar sands amidst the viral coverage of the ALS Ice Bucket Challenge (Canada Press 2014).

In December 2015, DiCaprio made headlines again, this time for comments made during an interview with *Variety* magazine. In the interview DiCaprio was reflecting on the weather during his time filming of the Oscar-winning movie *The Revenant* just outside Calgary, Alberta. DiCaprio remarked:

I've never experienced something so first-hand that was so dramatic. You see the fragility of nature and how easily things can be completely transformed with just a few degrees difference. It's terrifying, and it's what people are talking about all over the world. And it's simply just going to get worse.

We were in Calgary and the locals were saying, 'This has never happened in our province ever.' We would come and there would be eight feet of snow, and then all of a sudden a warm gust of wind would come (Tapley 2015).

What DiCaprio actually experienced was a Chinook; a weather phenomenon which occurs when warm winds blow in from the eastern slopes of the Rocky Mountains causing rapid snow melt as well as temperature swings of up to 20 °C (CBC News 2013). Far from being unprecedented, as DiCaprio claimed—it is not uncommon in Calgary—this very public lack of local knowledge was immediately jumped upon with local tabloid the *Calgary Sun* running a front page headline "LOAD OF DICRAPRIO" using different sized bold black and red capital letters so that "load of crap" clearly stood out. Meanwhile, oil sands industry supporters, Albertan politicians and critics ridiculed him for conflating a localized weather phenomenon with "terrifying" climate change (Nerman 2015). Critics used DiCaprio's mistake not only to engage in name calling (see Levant 2014) but to delegitimize his claims of negative environmental impacts and consequences of oil sands development arguing that if he did not understand the difference between a Chinook and climate change, how could he understand the impacts of oil sands development or even global warming? These critiques were frequently delivered with vitriol and disdain. As with James Cameron and Neil Young, critics also sought to emphasize the hypocrisy and disconnect between DiCaprio's carbon-heavy lifestyle as an A-list Hollywood celebrity and his call to keep carbon in the ground.

DiCaprio's mistake demonstrates that while celebrity endorsement can provide a valuable resource for environmental movements in their pursuit of publicity and legitimacy for their arguments, it is also a risk, with ill-informed or ill-conceived remarks providing ammunition for critics that may result in further polarizing publics.

## Conclusions

Taking protest against the tar sands in Canada as a case study, this chapter has explored some of the ways that environmental movements have set out to navigate a media saturated political environment and secure both visibility and legitimacy for their claims. Over a third (36.5%, n = 35) of the protest events mounted in the decade covered by my study employed a logic of physically "bearing witness" by demonstrating outside centres of political power or key sites of extraction. As noted, such interventions frequently employed playful and spectacular images, slogans and visuals which had been purposefully designed to capture public attention through both social and legacy media. However, as we have also seen, a larger number of events (40.5%, n = 39) involved the participation of celebrities raising questions about their increasing role in underwriting, orienting, facilitating or constraining political action.

More broadly, there is also a need to acknowledge the logic of celebrity as a mobilising logic of protest together with the logic of numbers, the logic of damage and the logic of bearing witness. To recognize the logic of celebrity is to acknowledge that the politics of late modernity are a politics of celebrity and media (Street 2004, 2012; Wheeler 2013). It is to recognize that celebrity is something that may be manufactured, refined, crafted, honed and exploited. As we saw in the cases of Neve Campbell and James Cameron, visits to the tar sands by famous individuals, particularly those drawn from the film and music industries, can amplify and transform the practice of bearing witnessing by propelling issues onto a popular news agenda in which the activities of celebrities have come to play a central role. As Neil Young's remarks confirm, celebrities are well aware that they possess cultural capital that may be deployed to support social justice campaigns. Reciprocally, as mentioned earlier, eNGOs have long been aware of the mobilizing potential of celebrity endorsements and the potential for their off-the-cuff remarks about issues to be quickly picked up, creating an enlarged space for discussion as with Neil Young's comparison of the devastated landscape around Fort McMurray with Hiroshima.

At the same time, as we saw with wave of orchestrated derision that greeted Leonardo DiCaprio's comparison of accelerating climate change with the familiar localised sharp variations in temperature caused by a Chinook, poorly informed comments can undermine a celebrity's credibility and reinforce doubts about their sincerity in defending environmental and minority rights carried by coverage of their lavish, and carbon dependent, life styles. The resulting public attention may weaken campaigning by deflecting attention away from core issues to focus on the character and motivations of the individual celebrity.

As long as the potential gains in increased visibility and legitimacy for campaigns are seen to outweigh the risks however, involving celebrities in activist media strategies will continue in tandem with the older established protest logics of numbers and witnessing, generating new possibilities for securing public attention and support for claims and proposals at the centre of campaigns. Exploring ways to maximise these possibilities in a popular news environment increasingly organised around personalisation and the instant viral distribution of images and comment on social media is a now an urgent challenge.

## NOTES

1. This research was funded by an Insight Development Grant from the Social Sciences and Humanities Research Council.
2. Initial media reports stated that 500 ducks had died yet the number was later revealed to be much higher with 1,606 ducks dying. Syncrude was fined $3 Million dollars by the Alberta government for the deaths (Wingrove 2010b).
3. Pictures of the event may be seen here: https://www.flickr.com/photos/tarsandsaction/albums/72157627473871226.
4. Video of the event may be seen here: https://www.youtube.com/watch?v=JBlrDzV_a98 and here https://www.youtube.com/watch?v=MyGc1F5niVk.

# CHAPTER 12

# Nothing but *Truthiness*: Public Discourses on the Adani Carmichael Mine in Australia

## Benedetta Brevini and Terry Woronov

Senator Waters, I am sure it has not escaped your notice that some forms of coal are cleaner than others and the coal that will be mined from the Carmichael mine by Adani is some of the cleanest coal in the world. So the consequence of the development of the Carmichael mine by Adani, the export of that relatively clean coal to India and its use in a new generation of Indian electricity generation to replace the pollutant biomass upon which those people rely at the moment will be to produce a much cleaner energy outcome (Senator Brandis questioned at Senate, 23 November 2015).

'Post-truth' has been celebrated by the Oxford and Maquarie Dictionaries as the 2016 Word of the Year, after dominating media and political discourses during the American election campaign and the Brexit debate in the UK. Defined as "relating to or denoting circumstances in which objective facts are less influential in shaping public opinion than appeals to emotion and per-

B. Brevini · T. Woronov (✉)
University of Sydney, Sydney, Australia
e-mail: terry.woronov@sydney.edu.au

B. Brevini
e-mail: benedetta.brevini@sydney.edu.au

© The Author(s) 2017
B. Brevini and G. Murdock (eds.), *Carbon Capitalism and Communication*,
Palgrave Studies in Media and Environmental Communication,
DOI 10.1007/978-3-319-57876-7_12

sonal belief" (Oxford Dictionary 2016), post-truth politics has thus been identified as a hallmark of the current era in the US and UK, which has had the effect of downplaying the ways that forms of political communication and spin that favour feelings and emotions over policy are spreading globally.

To demonstrate that post-truth politics and 'truthiness' are not only American/British phenomena, we focus here on the way that politicians and the media in Australia have debated the establishment of the one of the biggest coal mines in the world, the Adani Carmichael mine in central Queensland (Amos and Swann 2015; Taylor and Meinshausen 2014). We suggest that post-truth politics are not merely a replacement of 'truth' with 'lies,' but a complex, overlapping set of "discursive manipulations" (Carvalho 2007) that work together to produce very particular political effects. We begin with a brief discussion of the concept of 'truthiness' and provide a further theoretical elaboration. Then, drawing upon discourse generated in both the Australian Federal and Queensland State Parliaments, as well as reporting on parliamentary debates in the Australian media, we argue that close analysis indicates that more is going on here than simply a bitter partisan argument about the future of coal mining in Queensland and Australia. Facts and facticity were far less important in the argument in favour of building the Carmichael mine than regimes of 'truthiness,' where mine proponents generated 'affectively legitimated facts' (Gilbert 2016) about the mine and the environment.

We begin by introducing the controversial Adani Carmichael mine in Queensland.

## ADANI IN QUEENSLAND, AUSTRALIA: FACTS AND FICTION

In November 2010, the Indian energy company Adani Mining Pty Ltd began the process of seeking approval to build a massive greenfield coal mine in the Galilee Basin, a remote area of north-central Queensland, Australia. If built, the proposed Carmichael Coal Mine would be the largest in Australia and one of the largest in the world, producing 60 million tonnes of coking coal per year (Taylor and Meinshausen 2014) targeted for export to India's coal-fired electricity generating plants. If approved, this single mine would be a significant contributor to global climate change, producing 4.7 billion tonnes of greenhouse gas emissions, far above the 0.5% of the world carbon budget for limiting warming to 2 °C (Amos and Swann 2015; Taylor and Meinshausen 2014). In a recent joint report to the Queensland Land Court, two experts reporting on the carbon emissions of Carmichael's output warned:

Whilst the burning of the coal would not fall within Australia's national greenhouse accounts, the magnitude of the annual emissions associated with the burning of the coal would be equivalent to approximately three times Australia's annual emissions reduction target of 5% below 2000 levels by 2020 (Taylor and Meinshausen 2014, p. 10).

Because of its gigantic scale, enormous costs, and environmental impact, including the myriad ways pollution from the mine poses a significant threat to the already endangered Great Barrier Reef World Heritage Area, the Carmichael mine has been challenged in multiple court cases (Taylor and Meinshausen 2014). Political battles over regulating, approving and financing the mine, and the greenfield mining of the entire remote Galilee Basin, have been fought in the halls of both the federal and Queensland parliaments, and in the media across Australia.

How can a coal mine be subject to a regime of 'truthiness'? A proposal to build a greenfield megamine would appear to be an example of political facticity: economic, geologic, environmental and other related facts about mining and fossil fuel development should be arrayed to build arguments for or against. In the following, after reviewing the concept of post-truth and truthiness, we will demonstrate why official discourses around the Adani mine in Queensland and at the Federal level instead constitute an example of 'truthiness'.

## 'Truthiness' and Facts

The question of how to address the issue of 'post-truth' in politics and particularly in political campaigns has begun to draw attention from scholars in a range of disciplines, including media studies, communication and cultural studies (Gilbert 2016; Hannan 2016; Harsin 2015). Many of these authors have reflected on the usefulness of the word 'truthiness,' a term invented by (fake media personality) Stephen Colbert in 2005. As Gilbert (2016) notes, 'truthiness' is a potentially important concept for understanding contemporary political discourse, for it conveys the emotional quality of perceived realities, which, he argues, "are derived from passionate preferences rather than scientific, logical or even journalistic certainties" (ibid., p. 96). Citing Brian Massumi, he sees the language of politics now couched in 'a logic of "gut feelings" and "affectively legitimated facts"' (ibid., p. 96).

'Truth' in current regimes of truthiness is therefore not just facticity. Truthiness organizes affect, shaping judgments based on a pervasive sense that "all unwanted facts are political, uncertain and equally debatable" (Banning 2009, p. 293). Truthiness "aligns with a confused field of truth claims that alternates between false impressions and willful trades of complicated truths for constructed lies" (Gilbert 2016, p. 96). Public knowledge thus becomes "as much a matter of emotional expression as it is a collection of facts" (Gilbert 2016, p. 97).

Drawing on these theoretical elaborations, we suggest that when the following three traits characterize any particular political discourse, it may be labeled a regime of truthiness. First, arguments should follow "gut feelings" and emotional qualities. Second, it renders any 'inconvenient' or politically undesirable facts as debatable, doubtful or mere political ideology. Third, it relies on passionate political biases that are not based in scientific facts, logical argument, or empirical data.

To demonstrate this point, we look to both media and official discourses by Australian policy-makers on Adani's Carmichael Coal Mine[1] to investigate the extent to which a regime of truthiness dominates the debate on coal mining in Queensland.

## The Logic of 'Gut Feelings'

In the speeches and debates by Australian politicians in support of the Carmichael mine, the logic of affect and 'gut feelings' has emerged as a key element in public and political discourse.

One of the champions of this form of argument was the former Australian Prime Minister, Tony Abbott, who faced an onslaught of criticism from environmentalists after declaring that "coal is good for humanity, coal is good for prosperity, coal is an essential part of our economic future, here in Australia, and right around the world" and that it is Australia's "destiny" to "bring affordable energy to the world" (SMH 2014).

The passion with which Abbott and his fellow politicians defend coal mining cannot be solely explained as neoliberal ideology or corporate capture. The dominant discourses focusing on the 'destiny' or the importance of coal for humanity not only flies in the face of scientific facts, but adds an almost a religious connotation to the claims of coal's benefit to the nation and to mankind. In embracing a language based on facts that are 'affectively legitimated,' the former Prime Minister also relies upon a vision of mining as a foundational tenet in Australian culture, one that holds mining as central to Australian identity and way of life (Blewett 2012).

The affective argument used by politicians is that Australians are destined to exploit their island continent's rich mineral resources. Digging down into the earth makes the present Australian way of life and its continued future possible (Blainey 2001). In a 2009 *Quarterly Essay*, energy expert Guy Pearse described "the sacred place of mining and related industries in Australia today," a national imaginary he called "quarry vision" (ibid., p 4). Coal, he argues, is imagined to be part of Australia's "national competitive advantage endowed by providence", leading to a "perfect harmony on the importance of the quarry... From every direction Australians are told that their current and future prosperity depends on what we dig, drill and smelt for the world" (ibid., p. 1). An entire affective discourse based on the 'feeling' that mining is essential to Australian identity and Australian prosperity has been erected on this foundation.

This foundation is based on emotional qualities rather than empirical evidence. A comprehensive study by the progressive think tank The Australia Institute has demonstrated that Australia's economy would be barely affected by a moratorium on approval of new coal mines and mine expansions (Denniss et al. 2016).

Alongside the narrative of coal as national saviour, other 'gut feelings' arguments were used to promote the Carmichael mine. As evidence of Australian minerals' special status, unlike coal mined elsewhere in the world, Australia's coal was presented as assisting the world in battling environmental problems, rather than exacerbating them.

One excellent example of this reasoning is this quote, again from former prime minister Abbott during his visit to India in September 2014, when he once again praised the Carmichael project:

> It's one of the minor miracles of our time—that Australian coal could improve the lives of 100 million Indians, and it just goes to show what good that freer trade can do for the whole world (Prime Minister Transcript 2014).

Echoing arguments propounded by the chief executive of Adani, former Prime Minister Abbott pushed the moral argument that the mine could improve Indian living standards, alleviate poverty in India and also reduce carbon emissions. This last point was also put forward many times by Australian federal government representatives. In the words of Senator

George Brandis, Attorney-General for Australia, during a Question Period in the Senate, the mine would also help tackling climate change issues:

> I am sure it has not escaped your notice that some forms of coal are cleaner than others and the coal that will be mined from the Carmichael mine by Adani is some of the cleanest coal in the world. So the consequence of the development of the Carmichael mine by Adani, the export of that relatively clean coal to India and its use in a new generation of Indian electricity generation to replace the pollutant biomass upon which those people rely at the moment will be to produce a much cleaner energy outcome (Senator Brandis questioned at Senate, Hansard 2015).

Within this logic, the government argues that opening up the Southern Hemisphere's largest coalmine "will cut carbon pollution" (Hansard 2015).

## UNWANTED FACTS AS DEBATABLE, DOUBTFUL OR POLITICAL

A second component of 'truthiness' is the practice of deliberately presenting empirical facts as debatable, uncertain or political. This is accomplished in several different ways. One prominent tactic is the use of oxymorons to create contradictory messages. Kirsch and Benson (2010) derive their analysis of corporate oxymorons from George Orwell, whose book *1984* described how the state sought to make alternative thought impossible through the tactical appropriation and juxtaposition of key terms, rather than outright censorship. As they point out, in Orwell's use, some of these terms were euphemisms that meant almost the exact opposite of what they appeared to mean. Today, they note, corporations also rely on new oxymorons that rely on a figure/ground reversal that conceal the contradictions of capitalism (in this case the harms of the mine).

In the case of the Carmichael mine, proponents created new terms to conceal the harmful effects of the mine. One is the same tactic described by Kirsch (2010), the construction of a concept of 'sustainable mining.' As Kirsch describes, mining corporations globally have redefined the term 'sustainable' away from its original meaning, linked to environmentally sustainable projects, to now mean mining projects that provide employment (ibid.). In Australia, mine proponents were quick to use this redefinition of 'sustainability.' On 15 March 2016, Anthony Lynham, Queensland Minister for Development and Natural Resources, declared:

I will take confidence and intelligence over ignorance any day of the week. This government strongly supports the sustainable development of the Galilee Basin for the jobs and economic development that it will provide for regional Queensland (Lynham 2016).

Perhaps the most pernicious oxymoron used by mine supporters is the term 'clean coal,' which is increasingly used to defend constructing the megamine. Minister for the Environment Josh Frydenburg summarised this argument:

Australia's coal is predominantly low in sulphur and low in ash and can be used for these high-efficiency, low-emission power plants that we are seeing all around Asia which can reduce the carbon footprint by up to 40 per cent, and there are other new technology like carbon capture and storage which will be important to the future of clean coal (Frydenberg 2015).

This argument, repeated regularly in the halls of Parliament, is based on an analysis of Galilee Basic coal is the 'cleanest coal' in the world, which, if mined and burned, will actually reduce global warming.

In reality, as mentioned earlier, coal from the Galilee Basin is far from a clean solution. In their study of the impact of the Carmichael mine on climate action, Amos and Swann (2015) illustrated just how devastating those emissions will be. They note that the average annual emissions from burning the coal from the Carmichael mine—79 million tonnes of $CO_2$—is more than the annual emissions from Sri Lanka, more than Bangladesh with its population of 160 million, about the same as those from Malaysia and Austria, and only slightly less than the annual emissions from Vietnam. Compared to annual emissions from cities, the Carmichael mine's emissions will be three times the average annual emissions from New Delhi, double those from Tokyo, six times those of Amsterdam and 20% more than New York City (Amos and Swann 2015).

To counter the claim that Galilee Basin coal is 'clean,' further research by The Australia Institute points out that it 'is only 10% above the average quality of domestic Indian thermal coal in terms of energy content' because "the ash content of Carmichael coal is estimated to be 26%—more than double the average of 12% for Australian thermal coal" (Australia Institute 2016, p. 5). The Institute report reminds us that "any environmental impact comparison would [also] need to account for the requirement that this low energy, high ash thermal coal needs to be transported 5–10 times the distance of domestic Indian coal—adding additional greenhouse pollution related to transport" (Australia Institute 2016, p. 5).

Another 'truthiness' tactic used by pro-coal politicians to support their claims has been to re-frame the term 'activist' so that it connotes an enemy of both the Carmichael mine and the national interest. Government labelling of mine opponents moved beyond turning 'activist' into a negative term, with Parliamentarians calling members of green groups 'saboteurs,' 'vigilantes,' 'terrorists' and 'extremists,' and accusing them of waging warfare on Australia's economy. Far-right Queensland Senator George Christensen, for example, accused environmentalists of being enemy combatants who will damage the economy:

> [We must] end the ongoing warfare being waged by extreme green groups against major job-creating developments in the north. The recent setback with the Carmichael mine has alarmed many eager job seekers and business owners in the region who were looking at the opening up of the Galilee Basin and the expansion of the port of Abbot Point as a shot in the arm for our local economies (Christensen 2015).

This narrative casts environmentalists not only as economic enemies of Australia, but opposition to the mine as a form of terrorism. Senator Cristensen again, in the House of Parliament refereed to legal action to stop the mime as "an act of ecoterrorism" (Christensen 2016). He continued:

> Their lies, misinformation, slander and the frivolous legal action attacking a company for the sake of furthering an ideological cause can only be described as terrorism if you look at the criminal code (ibid.)

Australia is not the only place this discursive strategy of demonising environmentalists has been used. Recent scholarly inquiry into accusations of "eco-terrorism" (e.g., Potter 2011), however, has focused on whether or not environmental groups, such as Earth First and various Animal Liberation groups, are *really* terrorists, concluding that the vast majority of environmental activism cannot be genuinely labelled as 'terrorism' (Potter 2011). The accusations of "eco-terrorism" and "sabotage" related to anti-Carmichael mine activists have no foundation in fact whatsoever and are not linked to any actual illegal activities on the part of environmental groups.

## LIES: STATEMENTS THAT ARE NOT SCIENTIFIC, LOGICAL OR BASED ON FACTS

Parliamentarians and supporters of the Carmichael mine also rely on frankly mendacious statements to convince voters of their argument. Australia, once again, is by no means the only place this is occurring; as Giroux and Bhattacharya (2016, p. 2) note, "lying is part of an anti-politics of performance and showmanship that turns lies into political spectacles". The best examples related to the Carmichael mine are the claims of the number of jobs it will provide to the Queensland economy, where the employment situation is portrayed as desperate.

> The Adani Carmichael coalmine offers up to 10,000 new jobs, mainly in Queensland; $20 billion of investment in Australia; and power, to build the living standards of 100 million people in India (Landry 2015).

This figure of 10,000 new jobs has been repeated by virtually every mine supporter since 2014, including the Prime Minister, the Attorney General, and Liberal and National Party members of the Commonwealth and Queensland Parliaments.

Once built, the mine is not projected to offer anywhere near 10,000 jobs. The Adani Corporation's own economist predicted that the total number is closer to 1464: 'Adani's economist, Jerome Fahrer from ACIL Allen, found that Adani's mine and rail operations would employ around 1800 people directly and create around 1000 downstream jobs in "other services"' (Amos and Swann 2015, p. 1). Moreover, in testimony before a Senate Committee on how the Queensland government spends monies from the federal government, it was reported that mining is a small fraction of the Queensland economy, and is only viable due to significant subsidies from the state. Economist Roderick Campbell declared:

> The coal industry actually only accounts for 1.2 per cent of employment in Queensland. It contributes around four per cent of Queensland government revenue. If we stop and think about that for a second, that means that basically 99 per cent of Queenslanders do not work in the coal industry.... Last year, the Australia Institute released research showing that over six years the Queensland state government had put around $8 billion into spending that was primarily aimed at benefiting the coal industry and we estimate that

around $2 billion of that relates to developments at Abbot Point.... Deputy Premier Seeney said that the government was prepared to put in whatever it takes to make those projects viable (Campbell 2015).

Perhaps the most pernicious claim by mine proponents was summarized by Queensland Premier Anastazia Palaszczuk in the Queensland Assembly on 19 April 2016, when she stated: "Queensland taxpayers will not be funding any infrastructure for this project. Stringent conditions will be enforced to safeguard landholders" and traditional owners' interests' (Palaszczuk 2016). In actuality, to keep Queensland taxpayers from funding the mine's infrastructure, the burden will fall instead on the Commonwealth government via a proposed $1 billion loan to Adani for the rail lines from the mine to the coast (ABCnet 2016b).

Nor have the rights of the "traditional owners" of the mine site, a commonly-used euphemism to describe the Aboriginal groups who may have a Native Title claim to the land, been respected or upheld. Far from "safeguarding" these "traditional owners" interests,' the state, federal government and courts have so far denied all legal challenges from the Aboriginal people most effected by the Carmichael mine, denying their claims (ABCnet 2016a). The Wangan and Jagalingou (W-J) People, the traditional owners of the land on which the mine would be built, applied in 2004 for recognition of their native title (known as an Indigenous Land Use Agreement), which would give them the right to negotiate with the government in decisions related to the use of their land. Since then, the W-J People have filed multiple law suits intended to stop the authorisation of a Native Title permission to mine on their land. The W-J sued to reject Adani's mining lease on the grounds that the corporation had not received the approval of the full membership of the group. In a message to the Federal Court appealing the mining leases, W-J elder Adrian Burragubba said:

We said "no means no" and so we will continue to resist this damaging coal mine that will tear the heart out of our Country. The stakes are huge. In the spirit of our ancestors, we will continue to fight for justice until the project falls over. The decision of the Native Title Tribunal in April 2015 to allow the issuing of the mining leases by the Queensland government took away our right to free, prior and informed consent. It effectively allowed the government to override the decision that we made nearly two years ago to reject Adani's "deal".[2]

This appeal was dismissed in November 2016, and was re-appealed in February 2017, opposing the granting of mining leases to which the local group did not agree. Litigation continues in federal and Queensland court in late February 2017, complicated by a series of decisions elsewhere in Australia that called for Native Title Land Use Agreements to be approved by all members of Native groups (rather than a simple majority). In response, the Attorney General, George Brandis, has proposed an 'urgent bill' to change the Native Title Laws so that the Adani mining lease can be pushed through the courts in spite of objections by the Aboriginal stewards of the land (Guardian 13 February 2017).[3]

Another example of the deployment of 'lies' are the ways politicians have tried to deflect the effects the Carmichael mine will have on the neighbouring Great Barrier Reef. Although the Galilee Basin is inland from the coast, the mine will have devastating effects on the Reef, which is already under threat from climate change, acidification and agricultural run-off (Guardian 2016). Shipping the coal to India will require a massive expansion of the coastal coal port at Abott Point, including dredging of parts of the reef, and the amount of coal shipped through the treacherous waters of the Great Barrier Reef will drastically increase if the mine is approved. More generally, the mine's contributions to global $CO_2$ emissions will contribute to rising temperatures that are already killing the reef.

Politicians, however, have deflected these inconvenient facts by arguing that the mine's inland location absolves the Carmichael from any effects on the Great Barrier Reef; as Queensland Senator MacDonald explained

The Adani mine is some 300 kilometres from the coast. (...) Of course, 300 kilometres inland is nowhere near the Great Barrier Reef and all the protections in place to protect the Great Barrier Reef (MacDonald 2016).

## CONCLUSIONS

Post-truth may be 2016's 'word of the year,' but the use, meaning and implications of this changing form of public communication requires additional scrutiny. Our concern has been two-fold. The first is to highlight the ways that 'post truth' politics are currently extending beyond the US and UK to Australia, becoming a general feature of the arguments made by proponents of fossil fuels. The second is to detail how a regime of 'truthiness' is constructed in public debates.

Focusing on the political and media debates around the proposed greenfield Carmichael megamine in the Galilee Basin in central Queensland, we found examples of the three key strategies that construct 'truthiness' regimes.

First, mine proponents in Australia rely heavily on affective themes, building on pre-existing cultural tropes that cast mining as the nation's backbone, and coal as utterly essential to Australian safety, prosperity, and future. 'Gut feelings' rather than empirical facts drive arguments about Australia's destiny as a resource-extraction economy, allowing little space to mine opponents arguing that alternative economic paths may exist.

Second, following Benson and Kirsch's (2010) adaptation of George Orwell, we highlighted the specific use of oxymorons in public discourse, to show how newly minted terms such as 'sustainable mining,' 'clean coal' and 'terrorist activist' disarm opponents by generating euphemisms that mean almost exactly the opposite of what they appear to mean.

Third, employing these complex discourse narratives (Boykoff 2017) does not, however, preclude politicians and the sympathetic media from promulgating outright falsehoods. 'Truthiness' also depends on manipulation and endless repetition of 'facts' with no basis in reality to make political points. In the case of the Carmichael mine, two instances stand out. First, figures for job creation, economic growth and value to the Queensland economy have been wildly overblown, yet through sheer repetition false economic data has been taken as fact by the media and political classes. Second is insisting that the mine will cause no damage to the already endangered Great Barrier Reef.

Our main point in dissecting the arguments in favour of the Carmichael mine is to demonstrate the complexity of 'truthiness' regimes. None of these discursive forms—'gut feelings,' spin and politicization of unwanted facts or even outright lies—are enough on their own. Rather, these strategies overlap, intersect and reinforce each other, to create an overarching truthiness regime that posits new megamines as desireable, inevitable, and essential to maintaining Australia's national destiny.

The challenge for mine opponents is therefore formidable. Demanding that pundits and politicians replace 'lies' with 'truth' is not enough to topple this 'truthiness' regime. A more complex and multi-pronged approach will be necessary to convince the voting public that mining is not good for Australia, its economy, or the globe. As with climate change activists across the world, the battle against truthiness in Australia may be an uphill one.

## NOTES

1. We searched the Capital Monitor database for all references to the "Carmichael Mine" and the "Adani" corporation, the proposed mine's owners, between 01 January 2010 and 31 August 2016. Because the Galilee Basin, the site of the proposed mine, is in Queensland, we searched both the Federal and Queensland Parliamentary debates, legislation and law cases. We found 341 references to "Carmichael" or "Adani" between those dates in the Federal Parliamentary Hansard search, and 178 references in the Queensland state Hansard. References to the mine and the Indian conglomerate proposing the development were made by members representing all the major parties (Labor, Liberal, National, Green), during a time period when government switched from Labor (2010–2013) to a Liberal-National Coalition (September 2013–present) under four different prime ministers.
2. See: http://wanganjagalingou.com.au/traditional-owners-fight-on-appeal-carmichael-mine-federal-court-decision/.
3. Robertson, Joshua. "Coalition to Change Native Title Laws to Protect Mining and Agricultural Deals," *Guardian* online, 13 February 2017, https://www.theguardian.com/australia-news/2017/feb/13/coalition-to-change-native-title-laws-to-protect-mining-and-agriculture-deals.

# PART IV

# Communication and Campaigning:
# Oppositions and Refusals

**Fig. IV.1**  Get your head out of the tar sands

# CHAPTER 13

# The Anamorphic Politics of Climate Change

*Jodi Dean*

Politics in the Anthropocene is a matter of perspective. We can't look at climate change directly. Relying on multiple disparate measurements, we look for patterns and estimate probabilities. We see in parts; the melting ice caps, glaciers, and permafrost; the advancing deserts and diminishing coral reefs; and the disappearing coastlines and migrating species. Evidence becomes a matter of extremes, as extremes themselves become the evidence for an encroaching catastrophe that has already happened: the highest recorded temperatures; the 'hockey stick' model of predicted warming, sea-level rise and extinction. Once we see it—the "it" of climate change encapsulated into a data point or disastrous image—it's already too late. But, too late for what and for whom remains unsaid and unknowable. The challenge in this scenario becomes grappling with continuity. How can we conceive and wage the struggles already dividing the collectivity presumed in processes whose outcomes are estimated and predicted?

Climate change tethers us to a perspective that oscillates between the impossible and the inevitable, already and not yet, everywhere but not here, not quite. Slavoj Žižek reminds us that such oscillation indexes the 'too much or too little' of jouissance. For psychoanalysis, particularly in

J. Dean (✉)
Hobart and William Smith Colleges, Geneva, USA
e-mail: jdean@hws.edu

© The Author(s) 2017
B. Brevini and G. Murdock (eds.), *Carbon Capitalism and Communication*,
Palgrave Studies in Media and Environmental Communication,
DOI 10.1007/978-3-319-57876-7_13

163

Lacan's teaching, jouissance is a special substance, that intense pleasure-pain of enjoyment that makes life worth living and some things worth dying for. We will do anything to get what we think we will enjoy. We then discover after we get it that it wasn't what we really desired after all. Likewise, we try to discipline, regulate, and control enjoyment, only to find it emerging in another place. We get off even when we think we are trying not to. Jouissance is what we want but can't get and what we get that we don't want.

The Doomsday Clock is a symbolic clock face created in 1947 by the members of the Science and Security Board of the *Bulletin of the Atomic Scientists*. Its hands represent a countdown to possible global catastrophe. Initially made to refer to nuclear war, since 2007 the clock has been used to raise awareness for climate change.

Some use climate change as a vehicle for jouissance, for enjoying destruction, punishment and knowing. A current of left anthropocenic enjoyment circulates through evocations of unprecedented, unthinkable catastrophe: the end of the world, the end of the human species and the end of civilization. Theorists embrace extinction, focus on deep time and displace a politics of the people onto the agency of things. Postmodern Augustinians announce the guilt or hypocrisy of the entire human species. Hubris is humanity's, all of humanity's, downfall. Philosophers and cultural critics take on the authoritative rhetoric of geoscientists and evolutionary biologists. Those of us who follow the reports of emissions, extreme weather and failed states enjoy being in the know. We can't do anything about climate change, but this lets us off the hook when we stop trying.

Getting to name our new era, marking our impact as the 'Anthropocene', provides a compensatory charge—hey, we changed the world after all. Even better than coming up with a name for our era is the jouissance that comes from getting to judge everyone else for their self-absorbed consumerist pleasures—why didn't you change when you should have? Anticipatory Cassandras, we watch from within our melancholic 'pre-loss', to use Naomi Klein's term, comforted by the fantasy of our future capacity to say we knew it all along. We told you so. Your capitalism, instrumental reason or Cartesian dualism killed us all. Or so we fantasize, screening out the unequal distribution of the effects of warming —Russia doesn't worry about it as much as, say, Bangladesh.

The perfect storm of planetary catastrophe, species condemnation and paralyzed incapacity allows the Left a form of jouissance that ongoing deprivation, responsibility and struggle do not allow. Overlooked as too human, these products and conditions of capitalism's own continuity can be dismissed as not mattering, as immaterial. Organized political movement appears somehow outmoded, its enduring necessity dispersed into individuated ethico-spiritual orientations on a cosmos integrated over eons. This left anthropocenic enjoyment of destruction, punishment and knowing circulates in the same loop as capitalist enjoyment of expenditure, accumulation and waste, an enjoyment furthered by fossil fuels, but not reducible to them. Left anthropocenic enjoyment thrives on the disaster that capitalist enjoyment produces. In this circuit, captivation in enjoyment fuels the exploitation, expropriation and extraction driving the capitalist system: more, more, more; endless circulation, dispossession, destruction and accumulation; ceaseless, limitless death. Incapacitated by magnitude, boggled by scale, the Left gets off on moralism, complexity and disaster—even as the politics of a capitalist class determined to profit from catastrophe continues.

The circulation of left anthropocenic enjoyment through capitalist currents manifests in a diminished capacity for imagining human subjectivity. Even as things, objects, actants and the nonhuman engage in a wide array of lively pursuits, the anthropocenic perspective seems to confine humans to three roles: observers, victims and survivors. Observers are the scientists, their own depression and loss now itself a subgenre of climate writing. Scientists measure and track, but can't do anything about the unfolding catastrophe—action is for others. Observers also appear as the rest of us as moral audience, enjoined to awareness of human-nonhuman entanglements and the agency of microbes. In this vein, our awareness matters not just as an opportunity for spiritual development but also because multiple instances of individuated moral and aesthetic appreciation of fragility and the limits of human agency could potentially converge, seemingly without division and struggle. When the scale is anthropocenic, the details of political organization fall away in favor of the plurality of self-organizing systems. The second role, victims, points to islanders and refugees, those left with nothing but their own mobility. They are, again, shorn of political subjectivity, dwarfed by myriad other extinctions, and reduced to so much lively matter. The third role is as survivors. Survivors are the heroes of popular culture's dystopic futures, the exceptional and strong concentrations of singular capacity that continue the frontiersmanship and entrepreneurial individualism that the US uses to deny collective responsibility for inequality. I should add here that Klein's most significant contribution with *This Changes Everything* is her provision of the new, active and collective figure of 'Blockadia'. As is well-known, Blockadia designates organized political struggles against fracking, drilling, pipelines, gas storage and other projects that extend the fossil fuel infrastructure when it should in fact be dismantled. With this definition, Klein breaks with the anthropocenic displacement of political action.

If fascination with climate change's anthropocenic knot of catastrophe, condemnation and paralysis lures the Left into the loop of capitalist enjoyment, an anamorphic gaze can help dislodge us. 'Anamorphosis' designates an image or object that seems distorted when we look at it head on, but that appears clearly from another perspective. A famous example is Hans Holbein's 1533 painting *The Ambassadors*, in which a skull in the painting appears as such only when seen from two diagonal angles; viewed directly, it's a nearly indistinguishable streak. Lacan emphasizes that anamorphosis demonstrates how the space of vision isn't reducible to mapped space but includes the point from which we see. Space can be distorted, depending on how we look at it. Apprehending what is significant, then, may require 'escaping the fascination of the picture' by adopting another perspective—a partial or partisan perspective, the perspective of a part. From this partisan perspective, the whole will not appear as a whole. It will appear with a hole. The perspective from which the hole appears is that of the subject, which is to say of the gap opened up by the shift to a partisan perspective.

When we try to grasp climate change directly, we end up confused and entrapped in distortions that fuel the reciprocal fantasies of planetary scale geoengineering and post-civilizational neo-primitivism. The immensity of the calamity of the changing climate—with attendant desertification, ocean acidification, and species loss—seemingly forces us into seeing all or nothing. If we don't grasp the issue in its enormity, we miss it entirely. In this vein, some theorists insist that the Anthropocene urgently requires us to develop a new ontology, new concepts, new verbs and entirely new ways of thinking. Yet I have my doubts; geologic time's exceeding of human time makes it indifferent even to a philosophy that includes the nonhuman. If there is a need, it is a human need implicated in politics and desire, that is to say, in power and its generation and deployment.

The demand for entirely new ways of thinking comes from those who accept as well as those who reject capitalism, science and technology. 'Big thinkers' in industry and economics join speculative realists and new materialists in encouraging innovation and disruption. Similarly, the emphasis on new forms of interdisciplinarity, on breaking down divisions within the sciences and between the sciences and the humanities isn't radical, but a move that has been pursued in other contexts. Modern environmentalism, as Ursula Heise observes, tried to 'drive home to scientists, politicians, and the population at large the urgency of developing a holistic understanding of ecological connectedness'.

The Macy Conferences that generated cybernetics and the efforts of the Rand Corporation and the Department of Defense to develop more flexible, soft and networked forms of welfare, as well as contemporary biotechnology, geotechnology and biomimicry, all echo the same impulse to interlink and merge.

The philosopher Frédéric Neyrat (2015) has subjected the 'goosphere' that results from this erasure of spacing to a scathing critique, implicating it in the intensification of global fears and anxieties: when everything is connected, everything is dangerous. Neyrat thus advocates an ecology of separation: the production of a 'distance within the interior of the socio-political situation' is the 'condition of possibility of real creative response to economic or ecological crisis'.

Approaching climate change anamorphically puts such an ecology of separation to work. We look for and produce gaps. Rather than trapped by our fascination with an (always illusory) anthropocenic whole, we cut across and through, finding and creating openings. We gain possibilities for collective action and strategic engagement.

Just as it inscribes a gap within the supposition of ecological connectedness, the anamorphic gaze likewise breaks with the spatial model juxtaposing the 'molar' and the 'molecular' popular with some readers of Deleuze and Guattari. Instead of valorizing one pole over the other (and the valued pole is nearly always the molecular, especially insofar as molecular is mapped onto the popular and the dispossessed rather than, say, the malignant and the self-absorbed), the idea of an anamorphic perspective on climate change rejects the pre-given and static scale of molar and molecular to attend to the perspective that reveals a hole, gap, or limit constitutive of desire and the subject of politics.

Here are some examples of approaching climate change from the side. In *Tropics of Chaos*, Christian Parenti (2011) emphasizes the 'catastrophic convergence' of poverty, violence and climate change. He draws out the uneven and unequal impacts of planetary warming on areas already devastated by capitalism, racism, colonialism and militarism. From this angle, policies aimed at redressing and reducing economic inequality can be seen as necessary for adapting to a changing climate. In a similar vein but on a different scale, activists focusing on pipeline and oil and gas storage projects target the fossil fuel industry as the infrastructure of climate change, the central component of global warming's means of reproduction. But instead of being examples of the politics of locality dominant in recent decades, infrastructure struggles pursue an anamorphic politics. They don't try to address the whole of the causes and effects of global warming. They approach it from the side of its infrastructural supports. The recent victory of the campaign against the Keystone Pipeline, as well as of the anti-fracking campaign in New York State, demonstrate ways that an anamorphic politics is helping dismantle the power of the oil and gas industry and produce a counterpower infrastructure.

The new movement to liberate museums and cultural institutions from the fossil fuel sector supplies a third set of examples, modeling a politics that breaks decisively with the melancholic catastrophism enjoyed by the anthropocenic Left. As the demonstrations at the Louvre accompanying the end of the Paris COP made clear, artists and activists have shifted their energy away from the promotion of general awareness and participation to concentrate instead on institutions as arrangements of power that might be redeployed against the oil and gas industry. Pushing for a fossil-free culture, an array of groups have aligned in a fight against the sector that supplies capitalism with its energy. They demonstrate how the battle over the political arrangement of a warming planet is in part a cultural battle, a struggle over who and what determines our imagining of our future and the future of our imagining.

In this vein, Liberate Tate works to free art from oil by pushing the Tate to drop the sponsorship of British Petroleum. For the past five years, the group has performed art interventions in Tate buildings as well as other UK arts institutions that support (and are supported by) BP. Actions include unauthorized performances such as Birthmark, from late November 2015. Liberate Tate activists occupied the 1840s gallery at the Tate Britain, tattooing each other with the number of $CO_2$ emissions in parts per million corresponding to the day they were born. Hidden Figures, from 2014, featured dozens of performers standing along the sides of a hundred-square-meter black cloth which they held chest high, raising and lowering in arches and waves. Taking place in Tate Modern's Turbine Hall, the performance pointed to Malevich's Black Square, part of an exhibit that opened the same summer that carbon concentrations exceeded four hundred parts per million, a fact parallel to and omitted from the exhibit, much like BP's—and by implication the Tate's—involvement in the climate crisis.

Hidden Figures invokes the Tate's release of the minutes of meetings from its ethics committee in the wake of numerous freedom of information requests. Black rectangles blocked out multiple sections of the released documents. Hidden Figures reproduced an enormous black square within the museum, placing the fact of redaction, hiding, and censorship at its center. As Liberate Tate explains, the redactions reveal a divide, a split between the ostensible public interest of the Tate and the private interest it seeks to protect.

Occupying this split via its demonstration of the museum's incorporation into BP's ecocidal infrastructure, Liberate Tate disrupts the flow of institutional power. Rather than fueling BP's efforts at reputation management, it makes the museum into a site of counterpower.

The Natural History Museum, the new project of the art, activist, and theory collective Not An Alternative (of which I am a member), similarly adopts an anamorphic politics. The Natural History Museum repurposes the generic form of the natural history museum as a set of institutionalized expectations, meanings, and practices that embody and transmit collective power. It puts display on display, transferring our attention to the infrastructures supporting what and how we see. The Natural History Museum's gaze is avowedly partisan, a political approach to climate change in the context of a museum culture that revels in its authoritative neutrality. Activating natural history museums' claim to serve the common, The Natural History Museum divides the sector from within: anyone tasked with science communication has to take a stand. Do they stand with collectivity and the common or with oligarchs, private property, and fossil fuels? Cultural institutions such as science and natural history museums come to appear in their role in climate change as sites of greenwashing and of emergent counterpower.

Operating as a pop-up people's museum, The Natural History Museum's exhibits and tours provide a counter-narrative that combats the influence oil and gas industry on science education. The Natural History Museum also serves as a platform for political organizing, the ostensibly neutral zone of the museum turned into a base camp against the fossil fuel sector. It moves beyond participatory art's creation of experiences and valuation of participation for its own sake to the building of divisive political power. In March 2014, The National History Museum released an open letter to museums of science and natural history signed by dozens of the world's top scientists, including several Nobel laureates. The letter urged museums to cut all ties with the fossil fuel industry and with funders of climate obfuscation. After its release, hundreds of scientists added their names. News of the letter appeared on the front pages of the *New York Times*, *Washington Post*, and *Los Angeles Times*, and featured in scores of publications, including the *Guardian*, *Forbes*, *Salon*, and the *Huffington Post*. Later that spring, The Natural History Museum delivered a petition with over 400,000 signatures to the Smithsonian Institute in Washington, DC demanding that the museum kick fossil fuel oligarch David Koch off its board.

The premise of Liberate Tate and Not An Alternative is that institutions matter as combined and intensified expressions of power. More than just the aggregation of individuals, they are individuals plus the force of their aggregation. Because institutions remain concentrations of authority that can be salvaged and put to use, it makes political sense to occupy rather than ignore or abandon them. We can repurpose trusted or taken-for-granted forms—a possibility precluded by the anthropocenic preoccupation with an imaginary whole figured in geologic time. Just as the museum is a site in the infrastructure of capitalist class power—with its donors and galas and named halls—so can it be a medium in the pro-duction of a counterpower infrastructure that challenges, shames, and dismantles the very class and sector that would use what is common for private benefit.

The movement to liberate museums and cultural institutions from fossil fuel interests does not try to present climate change directly or nature as a whole. Instead, it approaches the processes contributing to global warming as processes in which we are already implicated. We are within the systems and institutions the effects of which scientists measure and chart. And that the people as the collective subject of politics are in them means that they are not fully determined. There are gaps that we can hold open and force in one direction rather than another. In too many contemporary discussions of the Anthropocene, the organization of people—our institutions, systems and arrangements of power, production, and reproduction—appears only as a distortion. Everything is active except for us, we with no role other than that of observers, victims or lone survivors. In contrast with emphases on nonhumans, actants, and distributed agency, the strategic coming together of organized opposition to the fossil fuel sector points to the continued and indispensable role of collective power. Just as a class politics without ecology can support extractivism, so can an ecology without class struggle continue the assault on working people that has resulted in deindustrialization in parts of the North and West and hyperindustrial-ization in parts of the South and East (we might call such an ecology without class struggle 'green neoliberalism'). So we shouldn't undermine collective political power in the name of a moralistic horizontalism of humans and nonhumans. We should work to generate collective power and mobilize it in an emancipatory egalitarian direction, a direction incompatible with the continuation of capitalism and hence a direction necessarily partisan and divisive.

# CHAPTER 14

# Journalism, Climate Communication and Media Alternatives

*Robert A. Hackett and Shane Gunster*

To the extent that journalism influences public agendas and remains modernity's most important form of storytelling, what kind of journalism could help humanity address climate crisis with the necessary urgency and openness to radical change?

A review of academic, professional and NGO literature on 'best practices' in environmental journalism identified multiple aspects of traditional reporting methods that could be revamped, including training, variability of topics, range of information and sources, balance and objectivity, newsworthiness and storytelling methods (Bourassa et al. 2013). While many proposals associated with these themes are worthy, they are too often disconnected from each other, and too modest in scope to supersize audience engagement with climate politics. Moreover, they often run counter to the anti-environmental logics of commercial news media, linked to corporate ownership, financial and fossil fuel capital, and consumerist culture. In much of the global North, newsrooms are economically imploding under the weight of debt, technological change and conglomerate disinvestment in journalism.

In this chapter, we suggest several more encompassing options for climate journalism—the (re)framing of climate politics by environmental

R.A. Hackett (✉) · S. Gunster
Simon Fraser University, Burnaby, Canada
e-mail: hackett@sfu.ca

© The Author(s) 2017
B. Brevini and G. Murdock (eds.), *Carbon Capitalism and Communication*,
Palgrave Studies in Media and Environmental Communication,
DOI 10.1007/978-3-319-57876-7_14

social movement organizations and supportive journalists; the implementation of new integrated paradigms for doing journalism; and the development of vibrant alternative media.

## FRAMING CLIMATE CRISIS: LIMITS AND OPPORTUNITIES

It is probably impossible to tell a story without a frame—a sense of what the story *is*, that draws upon broader cultural narratives and themes, and that helps shape what the storyteller regards as relevant, what categories and descriptive terms she uses, and so on. Framing is inherently a process of selectivity—as the metaphor implies, some aspects of the world are foregrounded, others marginalized, still others excluded altogether. It is also an exercise of power—the capacity to define a political issue (including climate and energy policy) through the use of frames that resonate with publics and help them 'fix' its meaning, is halfway to winning the policy battle. And the resources to establish frames that are favourable to one's group or class interests are not evenly distributed throughout capitalist society.

We conducted interviews with alternative media journalists and environmental NGO advocates in Vancouver to identify frames that they regard as relatively effective in engaging publics with progressive climate politics. We first asked them for their perceptions of current news.

Most participants talked about the sheer paucity of coverage, especially remarkable given the magnitude and complexity of the problem, as well as the urgent need for strong, immediate action. 'Limited coverage is part of the issue', noted P.J. Partington, a Toronto-based climate policy analyst with the Pembina Institute, a clean energy Canadian think tank. 'And that has to do with a lot of things, not least dwindling resources... [and] fewer specialist reporters...on science and environment.' Many noted the incongruity between reporting on climate change and a commercial media system dedicated to the promotion of consumption and economic growth.

Many were especially frustrated by the inability and/or unwillingness of news media to draw connections between climate change and other issues. In the context of British Columbia and Canada, the most important missing link is between energy and climate. Kevin Washbrook, a prominent activist and a founding member of the Vancouver grassroots advocacy group Voters Taking Action Against Climate Change (VTACC), identified the single most important thing news media could do to increase public engagement with climate change: "Connect the dots". On the one hand, media coverage of energy issues only focuses upon their economic and political aspects, but rarely mentions climate change. On the other hand, news about climate change 'is always about a new scientific study or the negotiations happening somewhere else, and they never connect the dots between that and fossil fuel exports in B.C.'. Instead, Washbrook argued:

> Every time you talk about how urgent it is to take action to reduce emissions and avoid the two degree ceiling, they need to say, 'And these fossil fuel export projects will play a direct role in increasing emissions'. There is a fundamental connection between what we do here in B.C. and this global, abstract concept .... We need to stop talking about [energy and climate] as two separate things. The Kinder Morgan [pipeline] protests and the energy economy of Canada need to be seen as connected and having direct application for climate.

Making these linkages not only helps people better understand the causes of climate change, but more importantly alerts us to our potential (collective, political) agency: as Washbrook put it, 'we are on the front lines of the battle to avoid dangerous climate change ... [and that] creates an opportunity, but it also creates an obligation, because we actually can do something here in Metro Vancouver'.

For some, this inability and/or refusal to pursue these linkages vigorously—especially between climate change and the fossil fuel industry—reflects the political economy of commercial media in Canada, particularly British Columbia. Jamie Biggar, former executive director of Leadnow.ca —a progressive political advocacy organization which runs national, issue-based campaigns—explained that a key 'disciplining factor' among news media is 'not wanting to challenge the fossil fuel companies'. As a whole, he argued:

> traditional media ... are very much a part of the governing class and there is a broad consensus in the governing class about the necessity of the oil sands in particular, and fossil fuel development .... And I think you... worry that you sound like you are not a serious person if you challenge that .... [And] the fossil fuel companies exercise a fairly significant lever over most traditional media now, as they have become enormous media buyers, enormous ad buyers.

Paraphrasing the sentiments of Rafe Mair (a long time B.C. politician, radio host and environmental activist), Damien Gillis—a documentary filmmaker and co-founder (with Mair) of the alternative news site

Common Sense Canadian—argued that an overarching 'pro-big business' frame operated as a de facto form of censorship in corporate media: young journalists quickly learn that 'you are either going to be censored, or more likely...self-censored ... You are not going to get published as much, [or] promoted as fast, [or] get paid as much, you are not going to get as many opportunities, if you do not play the game, if you do not know the script'.

In the rare moments when climate change does break through to the front pages, our participants suggested that it is invariably filtered through the prism of conflict. Geoff Dembicki served as the lead sustainability writer for the independent Vancouver online news outlet *thetyee.ca* for several years and he has written extensively about the politics of climate, environment and energy over the past decade. Media coverage of climate change, he suggested, has increased as the confrontational politics associated with carbon infrastructure projects have intensified, giving news a familiar template through which to represent the issue.

> And while I think it's good that climate change is getting more attention, and protests against pieces of infrastructure are getting covered, and those critical voices are being brought into the mainstream media, the result is that the entire mainstream narrative around climate change is almost always defined by conflict. So it's one group fighting another, one country calling out another country for not achieving targets, and it results in this very pessimistic frame where it's hard to feel that anything you do can have a real impact.

For Dembicki, conflict narratives are not only a barrier to agency and efficacy, they also lock key stakeholders and constituencies into a polarizing message track which prevents them from communicating with the public (and each other) in a thoughtful and constructive manner. Instead, their core communications objective becomes supplying news media with content that can be easily slotted into conventional journalistic formulas and, consequently, will generate media attention. While these patterns may produce good copy, they ultimately marginalize and exclude important issues from public discourse.

The rhetorical inertia of conflict narratives, compounded by their easy and productive articulation with dominant routines and patterns of news production, helps create a self-reinforcing cycle in which this simple form of storytelling displaces more complicated and less predictable accounts of climate change. It fortifies the 'prevailing view in the mainstream media and among the public that any progress on climate change is going to be fought over bitterly and will be decided through conflict', which ends up producing a real 'blind spot' with respect to solutions that can emerge out of design, technology or policy innovation.

David Beers (2006), publisher of *thetyee.ca*, identified three different, but complementary approaches through which solutions-focused journalism can 'catalyze concrete positive change.'

The first is 'living the solution', in which individuals describe their own experiences with a particular form of behavioural, institutional or social and political change. Exemplary of this approach was *The 100-Mile Diet* in which Alisa Smith and James MacKinnon wrote about their attempt to live on a diet consisting only of food sourced within 100 miles of their home. The popularity of the online articles led to a best-selling book (Smith and MacKinnon 2007) and helped spawn a global movement devoted to dietary localization.

Second, journalists can investigate and publicize innovative, local, small-scale experiments which are often highly successful but largely invisible to the broader public. In these cases, the analytic focus becomes questions of scale, reproducibility and barriers: if a specific practice, technology or policy is so effective, how can it be applied more broadly, and what social, economic or political barriers are preventing such expansion?

Finally, Beers noted the importance of exploring solutions in other jurisdictions, which are often neglected due to the parochial sensibilities (and shrinking resources) of mainstream newsrooms. Learning about and from other places can shake up public acceptance of the status quo, and enliven political debate about the full range of choices available to citizens and governments. How, for example, has Norway managed the development and governance of its energy resources (and the profits from them) compared to Canada? What might Vancouver learn from the cycling policies and infrastructure of Portland, Copenhagen or Amsterdam? While much of this information already exists in reports from NGOs and academic studies, the creativity and expertise of journalists as storytellers, combined with their ethical and professional commitments as fact-checkers, can give their reporting upon solutions a credibility and rhetorical appeal which can get the public interested, engaged and excited.

Especially important for both Dembicki and Beers are the identification of points of possible compromise and consensus—'moveable parts'—which are otherwise obscured by ongoing conflict between different stakeholders. One of the most surprising revelations for Dembicki in his series *Greening the Oil Sands*, for example, was the fact that 'the oil industry and environmental groups both support a carbon price, but they've never come together on the same stage and said, 'Prime Minister Harper, this is something we all support'. And the reason for that is… because they're fighting so much [both groups] can never see where they have similar objectives'. For Beers (2006), this was 'a huge story. It's one of the most censored stories ever. So finally Bloomberg [news service] comes around and goes, 'Ah, this can't be true'; phones up everybody and finds out it is true, and then they did their own story …. [F]ine-grained and credible journalists can be seen as honest brokers of that conversation in ways that activists and highly invested NGOs cannot'.

Honest brokers? Conversation? By contrast, some of our environmental communicator respondents expressed a more radical understanding of conflict frames, and by implication, the imperatives of journalism for social change. Is an emphasis upon conflict inevitably corrosive for efficacy, agency and hope? While all of our participants were critical of the predictable and formulaic patterns of conflict which dominate conventional news, some argued that conflict narratives are an inescapable and, in fact, essential part of good climate change communication. And rather than distracting audiences from engaging with solutions, conflict stories which intensify polarization, cultivate and focus outrage and celebrate struggle can facilitate the transition from awareness and concern to political engagement and activism (Gunster 2017b).

Biggar explained that there are two kinds of archetypal narratives which have taken shape around climate politics. The first emphasizes the failure of politicians and traditional institutions to address climate change, represents stakeholders (industry, government, environmental groups, First Nations, etc.) as gridlocked, and positions the public as disgusted but helpless bystanders to dysfunctional processes. 'That is a very demotivating, disempowering story that leads to cynicism.' And it is the story of climate politics which tends to drive conventional news agendas (Gunster 2011). However, he noted, 'there is another story in which institutional leaders are somewhat secondary, and what is actually primary is a fight between global fossil-fuel companies and place-based, but global civil society'. This second story is not only a more accurate representation of the current state of climate politics,

> …it is also much more empowering, because in the second story what you talk about…is victories and defeats, but what you are highlighting is normal

people who are getting involved, often successfully, against enormous odds. And that is really inspiring. And it fits with people's zeitgeist of the times, which is that things are really wrong in ways that are hard to articulate, and the levers of control of our society seem more and more distant. So where are there people who are taking things into their own hands and being successful? That is a much more motivating conflict story.

Indeed, Biggar argued that a 'huge part' of public engagement in climate change 'is figuring out what is an accessible conflict that you can get people into in order to challenge and hopefully transform what is going on. And, of course, the major answer to that is opposing pipeline projects, and other forms of dirty energy projects, where there is a physical, concrete thing on the ground that you can literally, physically stand up against and have a whole bunch of levers for trying to stop'. In other words, rather than dismiss all stories about conflict as alienating, one can distinguish between the paralyzing, cynical conflict frames recycled by conventional news, and the accessible, generative and mobilizing conflict narratives favoured by social movements and activists.

This latter understanding of conflict frames, linked to a broader analysis of power inequalities and of social movements as the engines of progressive change in capitalist society, is expressed by, say, George Monbiot (2009) at the *Guardian*, urging millions of people to take to the streets against politicians' inaction; Naomi Klein, author of a best-selling book subtitled *Capitalism versus the Climate* (2014); and Bill McKibben, founder of 350. org which is spearheading the fossil fuel divestment campaign. McKibben (2012) famously attributed the environmental movement's policy impotence to a failure to identify and target the fossil fuel industry as a primary enemy.

In evaluating frames, we do not mean to suggest that any one frame is universally applicable. What 'works' is likely to vary with cultural and political context, event and timing, and audience. Moreover, the 'common ground' and 'conflict' oriented frames discussed above are not necessarily mutually exclusive; even the former approach can lead to fundamental critiques of existing policy and power structures. The point we make here is that, while radical frames are obviously at odds with a media system dominated by corporate ownership and commercial imperatives, even the more modest frames and approaches do not mesh easily with the established routines and professional ideologies of hegemonic journalism. Calling on journalists to be more self-reflexive about the frames they adopt, to take into account the consequences of their reportage, to consciously tell narratives in the service of supposedly extraneous values like sustainability, or to mobilize people in a particular political direction, runs afoul of the regime of objectivity, with its semi-positivist epistemology, its assumption that news simply reports the facts as they are, and its illusory notions of press neutrality and independence (Hackett and Zhao 1998). Pro-climate framing is more likely to take root in whole new ways of thinking and doing journalism that challenge the regime of objectivity—in short, different paradigms.

## CHALLENGER PARADIGMS

Broader than individual story frames, a journalism paradigm comprises integrated elements that usually include distinct philosophical and ethical grounding, an analysis of how media work, a set of methods and procedures. What paradigms, or "corrective journalisms" (Cottle 2009), might help make journalism more truthful, ethical and adequate to the tasks of climate crisis? One starting point is to look to the emerging discipline of environmental communication (Gunster 2017a). American scholar and environmentalist Robert Cox (2007, p. 15–16) recommends a crisis orientation. Environmental communication should enhance society's ability to respond appropriately to environmental signals for the benefit of human and environmental health. It should make relevant information and decision-making processes 'transparent and accessible' to the public while those affected by environmental threats 'should also have the resources and ability to participate in decisions affecting their individual or communities' health', a notion that resonates with the concept of climate justice. Moreover, environmental communicators could engage various groups to 'study, interact with, and share experiences of the natural world', and critically evaluate and expose communication practices that are 'constrained or suborned for harmful or unsustainable policies toward human communities and the natural world'.

Could these criteria be transposed to journalism? They could help span the divide between the 'objectivity' standards of conventional reporting, and the 'advocacy' work of alarmed citizens. Their adoption implies a recognition that journalism is an inherently political practice, that there are already established models of engaged or advocacy journalism, and that nevertheless certain precautions would be needed: for example, avoid evaluating journalism through the single-minded lens of its environmental consequences; keep in mind journalism's distinct capacity for an independent gaze, one that separates it from propaganda or social movements (Calcutt and Hammond 2011). Cox's criteria are also reasonably consistent with the recognized democratic functions of journalism in monitoring power, surveying the physical and social environment for threats to well-being, and facilitating inclusive societal conversation on matters of public importance (Hackett et al. 2013, p. 36).

Fortunately, educators, researchers and practitioners who want journalism to productively engage publics on climate politics, do not need to reinvent the wheel. Two relevant paradigms emerged during the 1990s and 2000s respectively—civic (or public) journalism in the US, and peace journalism, internationally. Each generated a theoretical rationale, methodological guidance, pedagogical tools, some empirical evaluation, many on-the-ground projects—and a challenge to conventional journalism. *Civic journalism* (CJ; also known as Public Journalism) was born out of a widespread sense of democratic "malaise" and a "disconnect" between American publics, on the one hand, and politicians and media, on the other. Its core premise is the ethical and practical requirement for journalism not simply to report elite statements or to reproduce official political agendas, but actively to help reinvigorate public life. Drawing on the theories of American pragmatist John Dewey and German philosopher Jurgen Habermas, newspaper editor Davis "Buzz" Merritt and New York University academic Jay Rosen mapped out this new model during the 1990s, in the context of declining newspaper circulations and growing popular distrust of the press (see e.g. Merritt 1995; Rosen 1991). In some typical cases, news organizations (especially newspapers in mid-sized American cities) actively sponsored public discussions, ranging from neighbourhood pizza parties to town hall meetings, published information to animate such deliberation, and reported on the outcome. The idea was to engage publics and encourage them to articulate their concerns and priorities, and thereby to help set political and media agendas. The experiments carried out under the rubric of civic journalism—an estimated 600 by 2002 (Rosenberry and St. John III 2010)—have received mixed assessments. Supporters tout improved deliberative processes for communities, improved civic skills and favourable responses from citizens, responsive policy outcomes, expanded number of civic organizations, and broader inclusion of citizen voices in the news (Friedland and Nichols 2002). Others are more skeptical about the spread and staying power of CJ within corporate media, or its influence on journalists' professional attitudes or popular civic participation (Hackett 2017a). Nevertheless, both its limitations and its successes could richly inform our understanding of journalism's capacity to stimulate public engagement with climate crisis.

A second paradigm that might also resonate with those alienated by conventional climate politics news is *Peace Journalism* (PJ). Briefly, as outlined by two of its leading exponents, this paradigm is an analytical method for evaluating reportage of conflicts; a set of practices and ethical norms that journalism could employ in order to improve itself; and a rallying call for change (Lynch and McGoldrick 2005). In sum, PJ's public philosophy "is when journalists make choices—of what stories to report and about how to report them—that create opportunities for society at large to consider and value non-violent responses to conflict" (ibid., p. 5).

PJ draws upon the insights of Conflict Analysis to look beyond the overt violence that is the stuff of conventional journalism, which is often tantamount to War Journalism. PJ calls attention to the context of Attitudes, Behaviour and Contradictions, and the need to identify a range of stakeholders broader than the 'two sides' engaged in violent confrontation. If War Journalism presents conflict as a tug-of-war between two parties in which one side's gain is the other's loss, PJ invites journalists to re-frame conflict as a cat's cradle of relationships between various stakeholders. It also calls on journalists to distinguish between stated demands and underlying needs and objectives; to move beyond a narrow range of official sources to include grassroots voices—particularly victims and peace-builders. PJ seeks to identify and attend to voices working for creative and non-violent solutions; to keep eyes open for ways of transforming and transcending the hardened lines of conflict, and to pay heed to aggression and casualties on all sides, avoiding demonizing language and the conflict-escalating trap of emphasizing 'our' victims and 'their' atrocities. PJ looks beyond the direct physical violence that is the focus of War Journalism, to include other forms of everyday violence that may underlie conflict situations: structural violence, the institutionalized barriers to human dignity and wellbeing, such as racism; and cultural violence, the glorification of battles, wars and military power (Hackett 2006).

Obviously, PJ and environmental communication have different purposes and different targets—violent conflict and ecological degradation, respectively. In contrast to PJ, climate crisis journalism may be more likely to encourage investigative or watchdog journalism, and to be less reluctant to identify those most responsible for global problems, like climate change. But there are important affinities too. They have a similar critique of the limitations of traditional 'objective' journalism; they both honour the importance of framing and self-reflexivity in media, and recognize that journalism is implicated in the events it reports. Both seek a broader range of sources and more attention to context, positive developments, long-term processes, social structure, and creative ideas for solutions. And preliminary evidence from experimental settings in Mexico, the Philippines, Australia and South Africa suggests that by contrast with conventional war reporting, PJ framing does generate a greater degree of empathy, hope and cognitive engagement with counter-hegemonic arguments vis-à-vis war propaganda (McGoldrick and Lynch 2014)—precisely the kind of impact climate justice communicators hope for.

CJ and especially PJ, however, have found it an uphill battle to gain traction within conventional media. The reasons have to do with professionalism, resources and power.

First, both paradigms challenge aspects of the regime of objectivity, particularly strong historically in the US, which emerged in tandem with the commercialization and corporatization of the press, partly as a means of legitimizing growing media concentration (Hackett and Zhao 1998). Civic journalism does not advocate support for particular parties or policy options, but it does call for the press to abandon its neutrality on one key question—is public life working? Similarly, PJ calls for neutrality with respect to the contending sides in a conflict, but also for an explicit commitment to rendering visible the options for nonviolent conflict resolution. Furthermore, narratives that contextualize conflict may be open to accusations of partisanship in determining the relevant context: in the case of the 2003 invasion of Iraq, should journalists emphasize Saddam Hussein's previous use of chemical weapons against the Kurds, or the failure of UN arms inspectors to find weapons of mass destruction? These two contextual themes were preferred respectively by supporters and opponents of the invasion. Both paradigms thus seek a journalism that self-reflexively intervenes in political reality.

Second, in an era when major multi-media conglomerates are disinvesting in journalism, both CJ and PJ methods require a much greater investment of organizational resources than the standard reporting of official statements and photo-ops. In the US, CJ has apparently stagnated after media chains like Knight-Ridder stopped seeding it financially. Skeptical observers had suggested it was supported by media companies only as a circulation-building strategy, not as a philosophical commitment to deliberative democracy. PJ faces even further challenges, not only the resources needed to report and contextualize conflicts in distant places and multiple venues, but the barriers of constructing 'peace' as a compelling and newsworthy narrative (Fawcett 2002).

Third, to varying degrees, CJ and PJ challenge the policy agendas and even the ideologies of dominant economic and political elites, and thus, the bureaucracy-and power-oriented "news net" (Tuchman 1978) of conventional media. CJ calls for an infusion of popular voices to modify or even challenge established political agendas. More radically, PJ explicitly calls for deconstructing disinformation and war propaganda by states and other powerful institutions.

In these respects—traditional professionalism, news routines, resources, power—climate crisis journalism faces similar barriers, and could draw lessons from how CJ and PJ have surmounted or them—or not. In the meantime, there is are venues where such new ways of doing journalism are being incubated and developed—precisely, alternative and independent media, like *thetyee.ca* in Vancouver.

## ALTERNATIVE MEDIA

Unlike Peace and Civic Journalism, alternative media comprise not so much a coherent paradigm as a residual category: media that are 'alternative' to whatever media dominate the cultural and political "mainstream". In at least one sense, alternative journalism within such media is inherently political, in that it is "always a reminder of what the dominant forces in society *are not providing, or are not able to provide*" (Forde 2011, p. 45; emphasis in original). The proliferation of labels indicates the vitality and ferment in both practice and scholarship: alterative, alternative, autonomous, citizens, community, critical, independent, native, participatory, social movement and tactical media. Notwithstanding such diversity, we suggest the following 'ideal typical' characteristics, many of which resonate with practice and scholarship in environmental communication:

- oppositional or counter-hegemonic content (alternative frames; coverage of issues, events, and perspectives that are marginalized or ignored by hegemonic media; criteria of newsworthiness that emphasize the threats that the established order poses to subaltern groups, rather than vice versa);
- participatory production processes; horizontal communication both within news organizations, and with readers and audiences—communicative relationships that both reduce the gap between producers

and users, and empower ordinary people to engage in public discourse;

- mobilization-orientation; a positive orientation towards progressive social change, and productive connections with (but not subordination to) social movements;
- engagement with communities, whether defined in terms of shared locale, or shared interests;
- independence from state and corporate control, and from commercial imperatives; individual or co-operative ownership;
- low degree of capitalization; often reliant on volunteer labour, grants and donations; conventional production values and standards of professionalism may be less important (Atton and Hamilton 2008).

Those characteristics mesh well with climate crisis journalism that would seek to both inform and mobilize counterpublics, engage local communities and challenge entrenched power. In a study of alternative media coverage of climate change in Vancouver, Gunster (2011) concluded that it offered much more hopeful, optimistic and engaged visions of climate politics than the cynical, pessimistic and largely spectatorial accounts which dominated conventional news. While alternative media were deeply critical of the spectacular failure of 'politics-as-usual' at the 2009 Copenhagen summit, they invited the public to respond with outrage and (collective) action rather than (individualized) despair and hopelessness. Informed by a more sophisticated and broadly sympathetic understanding and exploration of the multiplicity of climate activisms, alternative media (re)positioned political action and engagement as viable, meaningful and accessible forms of agency for those struggling to respond to climate change. Particularly important in inspiring hope, were stories of political success—concrete examples of civic activism, political struggle, innovative and effective public policy, and transformative change in communities, institutions and governments (Gunster 2012). Such portrayals can normalize political engagement as something that ordinary people actually do, remind publics that another world is possible, and thus:

> not only feed upon the hope, nourished by historical example and consciousness, that democratic pressure can compel [existing] institutions to behave differently, but also awaken the political imagination to the utopian prospect of inventing new institutions and even new forms of politics in response to environmental crisis. Such an expansion of the conceptual and

affective spaces for climate politics produces an orientation that is simultaneously more critical and pessimistic about the limits of existing structures
and practices, yet also more optimistic about the opportunities for collective
political agency and intervention (Gunster 2011: 492–493).

More recent research in Australia supports these conclusions. By contrast
with the corporate press, independent and alternative media (such as
*Crikey*, *New Matilda* and the *Guardian Australia*) are more likely to
enunciate a clear commitment to addressing climate change, to critique
complicit and complacent governments and industry, to offer solutions,
and to encourage grassroots political protest and action. Alternative media
exhibited a different sourcing pattern—virtually no climate change deniers,
more environmental groups, more impacted but marginalized voices such
as Small Island Developing States (Foxwell-Norton 2017). Such media
evinced the facilitative role of journalism, particularly a conscious mission
to improve the quality of public life and to promote active citizenship
(Christians et al. 2009: 126). This sometimes shaded into the radical role
of supporting social movements representing disenfranchised groups, and
advocating more fundamental social change to eliminate concentrations of
power (Forde 2017).

How can this kind of journalism be scaled up to reach beyond the
mediascape's margins, however vibrant they be? Our research suggests
potentially overlapping agendas between alternative media producers,
media reformers, and effective climate justice communication (Hackett
2017b). As suggested elsewhere in this book, democratic media reform
may be one of the most effective ways to create an enabling environment
for climate crisis mobilization. Think big. Think laterally.

**Acknowledgements**  This chapter largely derives from Robert A. Hackett, Susan
Forde, Shane Gunster, and Kerrie Foxwell-Norton, *Journalism and Climate Crisis:
Public Engagement, Media Alternatives* (Routledge, 2017). We are grateful to
Routledge for permission to publish.

# CHAPTER 15

# An Interview with Alan Rusbridger: Keep it in the Ground

## *Alan Rusbridger and Benedetta Brevini*

As former editor of *The Guardian*, Alan Rusbridger put the threat of climate change front and centre. He believes that real change can only follow from citizens informing themselves and applying pressure on politicians and corporations.

**BB: Many recent interventions, from the interview with Greenpeace CEO David Ritter in this volume to Klein's book, indicate that the current climate crisis has a lot to do with how capitalism operates. Do you agree with this view?**

AR: Well yes, I think it's clearly all to do with capitalism and consumption and whether they're sustainable and whether companies care enough about sustainability as an ethical or environment[al] or... commercial part of their thinking. A lot of people think that... the patterns of consumptions in the Western world are incredibly destructive, so yes.

**BB: Developing countries are also following a very similar pattern. I thought that the launch of the campaign ['Keep it in the Ground'] was also linked to this idea, that actually the way in which big corpo-**

A. Rusbridger
London, UK

B. Brevini (✉)
University of Sydney, Sydney, Australia
e-mail: benedetta.brevini@sydney.edu.au

© The Author(s) 2017
B. Brevini and G. Murdock (eds.), *Carbon Capitalism and Communication*,
Palgrave Studies in Media and Environmental Communication,
DOI 10.1007/978-3-319-57876-7_15

rations operate is damaging the environment. "Keep it in the ground" marks one of the boldest advocacy campaigns ever launched by a media institution and one of the greatest legacies for *The Guardian*. What triggered the decision to embark on such a campaign?

AR: I had just decided to stop editing... I thought, "What am I going to regret when I don't have a paper?" And I think one of things that I thought was, had I done enough while I had the megaphone, as it were, to do everything I could on climate change? It wasn't that we'd done a bad job. We'd had more environmental reporters than any other paper certainly, probably most papers in the world, and we had our own environmental website. But somehow nothing that any of us was doing seemed commensurate with the nature of the problem. So I just sat down and thought, "What can I do before I go that would have some kind of impact? A real impact?"

BB: Can you tell our readers in detail what has the campaign entailed in terms of coverage, the number of articles, the timeline and targets?

AR: We set to work in January and I left at the end of the May so it was quite concentrated... five months from conception to end. We packed quite a lot into those five months and it began with me getting together about thirty correspondents and editors, going much more broadly than those who had just been working on the environment. So it included people who worked on economics, security, and home affairs and immigration, and all kinds of people for whom climate change had been marginal to that subject. We didn't go into it thinking this is going to be a campaign... It sort of grew out of a couple of conversations to begin with and it was quite a complicated campaign... We decided we would target the good guys rather than the bad guys because we thought that was more likely to have an effect. We talked to people who do campaigning for a living and they said, "This is an incredibly complicated thing to explain in terms of how campaigns normally work." But that was the origin of it.

BB: Then there was a conversation with 350 Degrees, I imagine?

AR: Yeah, well I'd met with Bill McKibben of 350.org in Stockholm the previous November and he was the one who said, "Look, you're doing it all wrong. We know the science now. The point is that this has now gone beyond science and we're only going to solve this if we tackle politics and the economics, so you have to broaden the range of people who are actually writing about this stuff."

**BB: Would you say that one of the most relevant goals of the campaign was to see divestment of big foundations and corporations from fossil fuels, or else to trigger a serious and more informed debate about climate change and global warming?**

AR: Both. We did one to try and get people to divest ... and that it seemed to be an important thing. But in the act of doing something that noisy and unusual and persistent in a newspaper, we hoped [it] would make people sit up and think about something in a way that just writing more news stories wasn't necessarily going to do.

**BB: So far, more than a year after the campaign, are you happy with the results? Do you think you were expecting more?**

AR: I think the results are amazing... I think the place where it struck home was the city, where people hadn't thought about this whole 'standard assets' issue. It looked amazing when the Bank of England was interviewed by the BBC and used language that was almost identical to the language we had been using, and essentially he was waving a big red flag in the air and saying, "Okay, this is not going to be something that's going to affect your investment portfolio this month or maybe next month but at some point in the medium term these companies are not going to be allowed to burn the assets that they have, and that is going to have financial implications." And I think there were quite a lot of people who hadn't seen it in that light before who began to think differently about the issue afterwards.

**BB: And so, for example, you feel the Gates Foundation started listening more?**

AR: The Gates divested a very large sum of money out of Exxon. They did it quietly about a year later, but they did divest and it had to coincide with a time when oil prices were extremely low and that sort of cut both ways. The valuation of oil companies was looking rocky anyway so people were thinking, "Is this the greatest investment in the world?" But also, I think it wasn't great for the renewables that oil suddenly became so cheap.

**BB: As a media reformer and scholar, I see how the lack of action on climate change can also be due to failure of the media. Do you agree?**

AR: Yeah... I think it's part of the big question about the current financing... I mean, it's not just financing... this was the biggest problem facing the planet. Why was it that the coverage of this subject wasn't remotely commensurate with that? I think there are numerous reasons. I don't think it's one. I think it's part of the fault of the readers. They maybe feel fatalistic, they maybe feel threatened, they maybe feel they've read it before, they maybe feel there's nothing they can do about it. Probably people don't much like reading about it. That translates in the modern world to something that is very measurable. If you're editing a paper and you can see what the readers are reading and not reading... that could influence their decision. You think, "If they don't want to read about it, why are we writing about it?" I think that newspapers are [in poor] financial condition anyway. They're all laying off staff and that kind of expertise is not there any longer... I think it's almost for politicians to take the decisions that are needed in the absence of any debate, and if there's not debate in the media then how on earth can politicians start educating the public about what's likely to be needed?

**BB: *Post-truth* seems to have become the 'buzzword' to describe this year's political climate. Fake news increasingly dominates social media, while traditional media institutions struggle to find continuous and reliable sources of funding. In these times, where do you see the future of climate change coverage heading?**

AR: I think again it's a very complicated question, and I think it's going to involve a long conversation with the big West Coast media companies who, at the moment, are cleaning up in terms of revenues but not willing to take the responsibility for what they publish... the Googles and Facebooks as well are going to have to be made more accountable and more transparent. They're going to have to start behaving more like editorial companies and less like distribution companies. I think that's one part of it which will deal with—which should deal with—a lot of the issues that are rising up at the moment about the so called post-truth society. Media companies are going to have to, I think, be more concentrated on convincing people, on winning back trust. I think a lot of the media have combined entertainment with journalism and have not exercised the same standards of transparency that are expected in public life nowadays. That's something they're going to win back, that trust, in many ways. For instance, to the extent there's no such thing as truth... the mainstream media are as bad as anyone else. If you're running a media company today, how would you demonstrate that you have a higher standard of truthfulness than some of the other stuff is available out there? I think that has to be more transparent: your sources and your working and footnoting and linking and being open to challenge and correction and clarification. All these things that would make people think, "Ah yes, here is an institution that operates to higher standards."

**BB: I don't know if you came across an article that Emily Bell wrote a few months ago describing the way in which Facebook was almost going to sign a deal for which, if you wanted to access the _New York Times_, you were going to have to do it completely through Facebook. What do you think of the idea of placing a global levy on companies like Google and Facebook as part of a media reform campaign?**

AR: I think Emily is one of the most interesting people at the moment writing about all this stuff, and it's striking. You have people who are technologically sophisticated who can understand these issues, because the issues of how Facebook worked with algorithms to select what you see and what you don't see and the degree to which they supposedly feed you stuff that fits in with your existing interests, all that is intensely interesting, and at the moment almost nobody understands it. If Emily slightly understands it then she's ahead of most people.

**BB: Given the lack of action by policy makers on global warming, are you worried about the direction that the US might be taking in climate policy under Trump?**

AR: Yeah, of course, terribly worried. But at the moment you can't tell to what extent anything he said during the campaign is going to be mirrored by what he does in real life. But yes, I'm very worried.

# CHAPTER 16

# An Interview with David Ritter: Mobilising on Climate Change—The Experience of Greenpeace

## *David Ritter and Benedetta Brevini*

David Ritter returned to Australia to take up his role as CEO of Greenpeace Australia Pacific in 2012 after five years working in various senior campaigns positions with Greenpeace in London. He believes that the current mess of climate change policy in Australia is a consequence of the maladjustment of politics and political economy.

**BB: On the third page of the latest COP 21 Paris agreement is the acknowledgement that the new $CO_2$ target won't keep the global temperature rise below 2 °C, the level that was once set as the critical safe limit. So we will officially pass the 2 °C level, either in 2036 or in 2038. What do you think of the media coverage? Do you think that the media have done a good job in communicating adequately on the Paris Agreement?**

DR: There is no question that overall media coverage of global warming is inadequate in the face of the scale of the issue ... Global warming dwarfs

D. Ritter
Sydney, Australia

B. Brevini (✉)
University of Sydney, Sydney, Australia
e-mail: benedetta.brevini@sydney.edu.au

© The Author(s) 2017
B. Brevini and G. Murdock (eds.), *Carbon Capitalism and Communication*,
Palgrave Studies in Media and Environmental Communication,
DOI 10.1007/978-3-319-57876-7_16

193

almost every other issue that the media routinely reports on. In a sense, it is a kind of ultimate threat because everything depends on having a functioning climate and the kind of media that we have is not used to addressing issues of that nature. I was recently reading the book *The Great Derangement* by Amitav Ghosh about the failure of the novel as a form to properly respond to climate change because of the extent to which it is historically rooted in as set of historically situated bourgeois sensibilities. I think that probably, at least in relation to newspapers or old-fashioned television 'nightly news', that there are conventions of the form that are contributing to the failure to cover the issue adequately. It isn't 'just another issue'—it is existential. Then there is the political economy of the industry: you've got the problem of those benefiting from the current system or are in denial of the need for change in the position of being newspaper owners or editors. The peculiarly destructive role of some news sources like Fox in the US is obviously particularly well known. At the other end of the spectrum, there are some media outlets which really do their best—the *Guardian* for example, and the Black Inc publications. *Crikey* also deserves a special mention as does the work of particular journalists like Peter Hannam at Fairfax. For government outlets, like the ABC, while lots of incredibly important work goes on, I think there is a sense in which absurd and outlandish positions on the climate science have been given too much air-time, with the search for balance leading to some rather ridiculous false equivalencies.

All of that said, trying to get across the inherently complex truth about Paris – that on the one hand it's an incredibly significant threshold moment getting the deal, getting the level of commitment to the targets on the one hand, but on the other hand the actual national pledges are woefully insufficient to get us there – is inherently difficult, even for well intentioned media because it does not conform to a neat narrative. The two are in tension because one can be reduced to 'we are saved' and the other to 'we are doomed', so telling both of those messages at once was probably a subtlety that the media was always going to struggle with.

**BB: As a series of recent interventions have made clear, including Klein's interview in this volume and her book, the present climate crisis is rooted in the current organisation of capitalism and its dependence on carbon-based sources of energy. What is your view on the matter?**

DR: Yes, I would agree with the view that the crisis of disembedded capitalism and the crisis of climate change are linked, and in multiple ways... Now it is true that we would still have a climate problem if social democracy had prevailed against the Washington Consensus in the 1970s and we'd absolutely still have a climate problem had communism had prevailed in the Cold War. There is no doubt that state communism itself was both an environmental disaster and hardly respectful of scientific discourse (remember Lysenko!). However we do need the state to take a more strategic role in society and to be willing to plan and intervene to solve the great problem of climate change... So yes, I think being nostalgic for the social democratic and social liberal impulse for collective problem solving is apt and that Klein is right to point out that just as government has been at its weakest is a moment when we need government the most... Now where I think it's interesting is that it's arguable that the financial, the neoliberal response to climate change whereby you marketise everything—you create markets here, you create markets there, you create markets in natural forests and the rest of it—that probably reached a high point in and around Copenhagen. That neoliberal view of how to respond to the climate crisis among those who are part of that world but who accept science also seems to have been defeated and that itself is interesting, probably because it tells us something about the contest within capitalism. One way of reading 2009 is that it was the financial sector versus the extractive sector—and the financial sector lost. Arguably the financial sector was at the weakest it had been for years at that point, in the wake of the global financial crisis. It is also worth recalling, though, that real and existing 'neoliberalism' never really adheres to the theory of a smaller state. It's never really smaller government, it's always just re-regulation in the interests of the wealthy.

**BB: The media system is one crucial arena in which competing accounts of climate change struggle for visibility, credibility, and legitimacy. In the experience of Greenpeace, does the proliferation of online opportunities for communication work to the advantage of campaigning groups? Can you mention some success cases?**

DR: I can, but perhaps I should preface my remarks first by saying I absolutely agree that the mainstream media does convey the ideology of consumption: the way to make yourself happier is to buy things... Recently I have been quite influenced by reading Evgeny Morozov and looking both at continuities and disruptions in the impact of online communications and communities. Unquestionably we have seen a step change in the ability to organise fast and the ability to deliver critical mass around an issue. But at the same time some of these things are losing their power as they become more and more easily replicated. The first example I had with Greenpeace was where the first piece of video content I was associated with, and I'd only been working with the organisation for a few weeks, targeted a particular brand over links to deforestation in Indonesia. I think it got only a couple of hundred thousand views, but in those days that was considered remarkable —this was late 2007. On the other hand, having whatever we have now reached maybe approaching ten million people around the world signed up to Arctic Protection—was helpful in a much more positive way. I'm told that very quickly assembling many, many people around the world to digitally lobby John Kerry helped unite the beginnings of the conversation about a UN convention over law of the sea implementing agreement which may one day allow high seas marine reserves. It is now almost unthinkable that a key moment of social change would not be accompanied by mass social media activity, at least in a developed nation. So I think one can see many examples of this, of when social [media] does provide an opportunity. But we have also seen social media allowing the proliferation of untruth... I think it's a dynamic rather than being overwhelmingly a good thing.

**BB: You've had impressive international experience with civil/environmental groups. Do you think that the media in the UK was more or less hostile to environmental groups than the Australian media? Could you give us specific examples?**

DR: It's not simply the malevolent impact of the Australian or some of the Murdoch papers but the way in which that shifts what is regarded as 'balance'. So when you have the national broadcaster coming under pressure to have what is described as balance where one part of the see-saw, if you like, is occupied by someone who doesn't reside in reality because they don't believe science, that's not proper balance at all... I think things probably are more unbalanced in that sense in Australia because of the concentration of media ownership. I think also we do suffer from being a political geography that is quite diffuse, with some very isolated political centres in the outer capitals... Old fashioned anti-intellectualism also remains a problem in Australia. My perception is that people who work at universities in the UK aren't attacked in the way they are in Australia. For all the talk of innovation and all of the talk of wanting to be smarter and more clever and all of that, it never seems to result in being a little bit more thoughtful or wanting to actually honour our universities or honour our intellectuals.

**BB: That probably created a divided environment for this push against the environmental movement.**

DR: I think there is a vast constituency for the environment in Australia, but that the kind of language we use really matters. I don't talk to many people who hate trees! I don't talk to many people who don't want fish in the sea when they go fishing on the weekend with their kids. I don't talk to many people who don't want to be able to walk to work with air free of gunk. They don't want pollution. They don't want pollution in their kids' playground. They don't want pollution in wherever the local place is that they go and play on the weekend. People want green space. They want shade. I've never met an Australian that wasn't very proud of the wonderful animals in that live in this country or amazing places like the Great Barrier Reef and the Great Australian Bight. Now what you might find is that when you're talking to people you don't know, at a barbeque or in the shops or a children's birthday party, you might not find people who say, "Well look, I'm concerned with the latest biodiversity indicators and frankly I'm very disappointed in the way that our national indicators have been structured arising out of the COP talks," but you will find people talking about the amount of plastic in their kid's bedroom, or the rubbish on the beaches, or the fact that they feel like the country is out of control and politicians are not acting like grown-ups. You will hear people worrying about spending too much time in cars and not being happy that the air doesn't taste good. I think it's all linguistic tricks. You can write off policy wonks as people who are not connected with society, but I simply think it's absolute bullshit that the people of this country don't care deeply about clear air, clean water, thriving animals and the places we love.

**BB: Australia has the highest per capita emissions anywhere in the developed world and while the Great Barrier Reef is dying, Australia insists to open the largest coal mine anywhere in the world. How is this even possible? As a scholar, journalist and media reformer, I am convinced that the lack of action on climate change is partly due to an unprecedented media failure. Do you feel the same? What can be done about it?**

DR: I agree it's not just neoliberalism and it can be a mistake to use the label 'neoliberalism' as shorthand for a range of structural factors that underly the state of things... One of the things I like about Naomi's *This Changes Everything* is that she coins this term "extractivism" and I think we live in Australia in an "extractivist" political economy and that explains a lot about what happens around the place. Anything that does anything other than just get stuff out of the ground as fast as possible is in some way defined as being a 'nuisance', a 'roadblock', 'unrealistic', almost no matter who owns the mine or how much it's going to benefit. As we can see from the court case in which the Adani fellow was cross-examined on his job's figures, it seems you can almost just make stuff up. If it was just neoliberalism it would be different, but it's this "extractivist" political economy and that is rooted deep in our culture, even though it has always also been contested… Last year when I was in Canberra at Parliament House I had a few conversations which illustrated the point. At that stage we were calling for an inquiry into the power of influence in the fossil fuel/mining industry, and the problem that I had speaking to journalists about it was that they tended to say: "This isn't a story. This is something we already know." Effectively, people were saying 'come on mate, this is just how things are, we all know that'. I thought that said a lot. That the fact that people simply accept that power in this industry: the malign power it has within Australian democracy, despite the fact that it's by some measures really quite a marginal economic concern, despite climate, despite what coal pollution does even without climate, despite the fact that it's just such a negative vision of Australia and the Australian people: that destructive and negative idea that all we can do as a people is dig holes in the ground.

**BB: Given the failure of mainstream media to effectively convey the issues surrounding climate change, have you connected with any media reform groups in the UK to collaborate on environmental and social movements?**

DR: I've never been contacted, at least never in the sense of, "Hey, would you be interested in running a campaign around X issue?". But I certainly wouldn't rule it out as an issue that Greenpeace could campaign on for the reasons that we've outlined: that it's extremely difficult to shift mindsets around consumption or extractivism or anything else when one has the media acting like a propaganda machine for the status quo, which people do unwittingly. In other words, media reform could be perfectly within our organisational mission of securing an earth capable of nurturing life in all its diversity if it was seen as precondition to achieving that end. I have had private conversations with a number of significant figures in the area of media reform. What I can say is that I have felt that there has been a meeting of the minds about the relationship between the problems.

**BB: I'd like to close on a hopeful note. After a successful launch in Canada, the LEAP Manifesto has gained momentum in Australia. Do you think change is imminent?**

DR: There is no doubt that change is coming. People working together can achieve anything. LEAP is exciting and has inspired lots of hope and movement along with other projects like Next System and NEON. People are starting to come together in Australia too, in an organised way, in a collective project to secure a better future. We must come together, to share some analysis on what some of the foundational causes are that underpin economic, social and environmental problems, to build political community and to organise. Because whether you're talking about homelessness or you're talking about fracking... refugees, whatever the issue might be, there are certain common problems of power and structure that need to understood and challenged... And the democratic deficit manifested in the media is one of the basic structural issues that we have to face up to. Because hoping you're going to get a favourable article in an antagonistic newspaper with a declining readership or shackling your civic conversation to Facebook's latest algorithm change is not actually a plan for how you build a sustainable, thriving democracy. We need a plan. I retain an absolute conviction that we will make a great transition to a fairer, more just and sustainable society—but we will only achieve that great dawn through political collaboration, through shared determination and through sheer hard work.

# CHAPTER 17

# An Interview with Blair Palese: Green Campaigns—Challenges, Opportunities and 350.org

## Blair Palese and Benedetta Brevini

In Australia Blair Palese and 350.org are working to build a movement to stop new fossil fuels and rapidly reduce greenhouse emissions by countering the power of the fossil fuel industry with the power of people taking collective action.

**BB: According to some of the most recent and powerful interventions including the book by Naomi Klein, free market fundamentalism and capitalist globalization have generated the current climate crisis. Do you agree with this view?**

BP: There's no doubt that the notion of endless growth—profiting from the use of every resource you can get your hands on regardless of consequences—has led to the current climate change crisis. Run-away capitalism and the burning of fossil fuels for energy is unsustainable and we need to change how we interact with our planet if we are to survive. The issue of climate change

B. Palese (✉)
Sydney, Australia
e-mail: blair@350.org

B. Brevini
University of Sydney, Sydney, Australia
e-mail: benedetta.brevini@sydney.edu.au

© The Author(s) 2017
B. Brevini and G. Murdock (eds.), *Carbon Capitalism and Communication*,
Palgrave Studies in Media and Environmental Communication,
DOI 10.1007/978-3-319-57876-7_17

offers us a unique opportunity. If we can solve this as a world, it means we can take on future challenges with new abilities and a global awareness of our impact on the planet. But it starts now and every country needs to immediately begin the transition to to clean, renewable energy.

**BB: Klein, in an interview for this volume, highlights possible solutions for the lack of serious climate policy action. She mentions new alliances and a growing global movement of indigenous people to defend their lands. However, Donald Trump's election as President of the United States looks likely to strengthen the hand of both the fossil fuel business interests and climate change sceptics. His pick to lead the Environmental Protection Agency for example remains sceptical about the need for urgent action on climate change. How should we respond?**

BP: Let's make no mistake, the Trump administration and global trend toward inward-looking nationalism poses a strong threat to our climate and other social justice issues. The next few years will challenge us, but when our governments and laws fail to act in our best interests we have a responsibility to step up and take action. To this end it is up to all of us to support local action to forward progressive causes, including climate change action. There is some inspiring action happening at the state and city level in many countries where citizens are taking action to put in place a rapid move toward renewable energy. Globally, indigenous peoples are standing up against the abuse of land and water that belongs to them and we have to find ways to become good allies to these causes. It won't be easy but buoyed by a clean energy industry that is now cheaper than fossil fuels and a growing global movement building political power for climate change action, we have to keep fighting during the Trump years.

**BB: Environmental groups have rapidly embraced social media. You are leading an organisation that runs climate-focused campaigns and, reading from the website, "runs projects and actions led from the bottom-up by people in 188 countries. Email is how 350 connects and reaches out to our support base." Do you think that social media has brought new possibilities for environmental groups? Could you give some examples? Do they also pose problems?**

BP: 350.org was founded globally with a very 'open source' approach and the use of digital outreach to bring people together in the most effective way possible. Our first video used only numbers and graphics so that it could be shown and used around the world in any language and culture. Using email and social media means we can respond quickly, organise globally and connect people with common battles even if they are worlds apart. Our first big day of action that saw people putting the number 350 in iconic places—the Great Pyramids, the Eiffel Tower, the Sydney Opera House, etc.—was organised with a website and an event location map that allowed anyone to organise their own event in the way that worked for them or find an event nearby. More recently, our Break Free events around the world saw people take part from the UK and Germany to the Philippines and Australia's Newcastle Harbour—the world's largest coal shipping terminal—to stop fossil fuels. We couldn't organise any of these events without these new digital channels. There is a risk of being tracked by authorities who want to stop protests from happening, clamp down on environmental organisation funding and discredit the call for the end of fossil fuels. Our challenge is to have a strong and organised movement in as many countries as possible so that we can support each other and learn as we go.

**BB: For one of the biggest climate marches in the world, you also developed sample tweets and a social media toolkit for the global climate march. Can you tell us more about this strategy?**

BP: Most people who are concerned about climate change want to take action but not everyone knows how or what they can do. 350.org, and the many partner organisations we work with around the world, tries hard to make it easy for people to get involved. By preparing social media information, artwork and flyers that are replicable and online resources so people can find local groups and events makes all the difference. Make it work for people and they will come!

**BB: Do you think that the new digital strategies by environmental groups can effectively counter mainstream commercial media that have perpetuated doubts and clearly failed to inform the public adequately on climate change? Can you give us an example from your own campaigns?**

BP: The mainstream media is fighting hard to keep readers, listeners and viewers. Free digital media makes it hard for mainstream media to keep a large enough paying base to operate in the way that they always have—with specialist and investigative journalists that have been so important as part of democracies around the world. 350 has worked closely with outlets like the *Guardian* to keep the kind of important coverage of climate change going despite these changes and in the face of media networks with vested interests. It's a David and Goliath fight but a critical one. Our co-founder Bill McKibben and 350.org board member Naomi Klein regularly contribute to mainstream media networks to try to ensure we can keep educating people about the latest science, the severity of climate change impacts we are seeing around the world and the many benefits of shifting to clean energy. Whether it's around our Break Free protests, ensuring important science is getting the attention it demands or calling out climate change deniers on their misinformation, we use all of our digital muscle to get the word out and provide opportunities for people to get involved on the issue.

**BB: Now, a piece of good news. The global movement to divest from fossil fuels has doubled in size since September 2015. Can you tell us a bit more about this global campaign in which 350.org is engaged?**

BP: The divestment of more than A$6 trillion (US$5.2 trillion) from fossil fuels is something that we in Australia and the 350.org team globally are particularly proud of. The campaign started off very much as a symbolic effort—much like the divestment effort during the Apartheid years. But much to our amazement, it has become all about money—and lots of it!—being moved from coal, oil and gas. By the end of 2016, 688 institutions and 58,399 individuals across 76 countries have committed to divest in some way. This includes the Norwegian Sovereign Fund, the Rockefeller Bros Fund, Stanford University and, here in Australia, 30 local councils, one local state government (the ACT), two universities and a number of investment funds. This is a real indicator that people are frustrated with government inaction and are taking their own steps to be part of the climate change solution.

350.org continues to work with funds, churches, universities and governments to divest from fossil fuels and this year we are campaigning to see the Nobel Foundation and the Louvre museum divest as well. Institutions everywhere have an important role to play not only to send the signal that moving from fossil fuels is the right thing to do but it's also the smart thing to do if you want to protect your investments.

**BB: After the good news—let's talk about Adani. Plans are alive to build the biggest coal mine in the southern hemisphere in the middle of Queensland—a mine that will wreck the already endangered Great Barrier Reef. Are you optimistic about the possibility to stop Adani, given the Federal and local government support it gained? What would it take to stop it?**

BP: The Adani mine is just one more example that Australia's current Turnbull government is more interested in governing for vested interests than for Australians. Pandering to fossil fuel interests for political gain, it has pushed the project and attacked any opposition to this mine—from traditional owners, farmers and environmentalists. However, I am positive that we can kill this project before it even begins. Currently, one of the largest coalitions of Australians ever assembled against a coal mine is mobilising to stop Adani from going ahead. We know that this mine is bad business—it would blow any chance for a safe climate future, undermining Australia's Paris pledge to keep warming below 2°, not to mention coming at a time that coal is in structural decline. By highlighting this fact and targeting the banks and other possible investors—getting them to rule out funding the project—we can ensure that despite government support that the project doesn't get the finance it desperately needs to go ahead.

**BB: How do you change the quite common view that climate change is a luxury concern for people who are very privileged? How do you build a climate justice movement that can bring together daily economic concerns as well as concerns about justice, jobs, services, health and the need for climate action?**

BP: There is no doubt that those who've done the least to cause the problem are feeling the worst impacts of climate change right now and will do in the future. You only have to look at sea level rise and extreme weather events in the Pacific islands and countries like Bangladesh and the Philippines or drought in Africa to know that climate justice is an essential part of the solutions we need. The developed world must play its part in helping the nations most vulnerable to the impacts as well as helping developing countries leapfrog fossil fuels straight to clean energy. Again, climate change offers us a unique opportunity to address the issue as a unified word. At 350.org, we've tried not to tell people in countries around the world how to campaign on climate change, but to ask them what they think works for them and how we can join them in the effort.

**BB: Mark Dowse argued in his 1995 book, Losing Ground, that the green movement needs to examine and criticize itself, or "it will become merely a clever marketing hook and even less relevant to the problems we face in the 21st century". How much has the environmental movement evolved since then?**

BP: Climate change is an incredibly difficult issue for people to get their heads around. Complex, yes, but more importantly, made confusing by media with vested interests questioning climate change science for decades —a tactic straight out of the tobacco playbook. Understanding that science is never 100% certain is not something most of us think much about, so it's easily exploited.

That said, those of us working on the issue do need to reach out more broadly to everyone, everywhere about the importance of the issue and ensure our movement spans the political, race, gender and age spectrum to be truly representative. It's worth remembering that the People's Climate Marches of 2015 saw more than 600,000 people take to the streets in 175 countries—by all accounts the biggest thing of its kind in history. Participants included union members, nurses, doctors, lawyers, business owners, students, refugees, members of the lesbian and gay community and people from virtually all ethnic groups. We've made real progress but there's certainly more to do.

350.org was founded in 2009 with an inclusive approach and we continue to live by author Naomi Klein's mantra: To change everything, it will take everyone. Our job now is not only to raise the alarm about climate change but to make clear the huge benefits we all get from solving this huge challenge. Our voices need to be loud and everywhere if we are to successfully stand up to fossil fuel dollars and influence. Knowing this, our focus now is to continue to grow and expand around the world so that the call to action can't be ignored.

# CHAPTER 18

# Conclusion: One Month in the Life of the Planet—Carbon Capitalism and the Struggle for the Commons

*Graham Murdock*

A series of publications and events in March 2017, when this volume was being prepared for production, provided comprehensive new evidence that the climate crisis was accelerating, underlined the growing strength of the struggles around it, and gave formal status to an alternative framework for thinking about a viable future built around a reassertion of the idea of the commons. All of these strands have long histories and deep roots but looking more closely at this one month in the life of the planet points up the central argument advanced by the chapters collected here; that understanding the operation of carbon capitalism and constructing alternatives to it must take the organisation of communication systems, as both arrays of material infrastructures and machines and contested cultural spaces, centrally into account.

G. Murdock (✉)
Department of Social Sciences, Loughborough University,
Loughborough, UK
e-mail: g.murdock@lboro.ac.uk

© The Author(s) 2017
B. Brevini and G. Murdock (eds.), *Carbon Capitalism and Communication*,
Palgrave Studies in Media and Environmental Communication,
DOI 10.1007/978-3-319-57876-7_18

207

## ACCELERATING CRISIS

On the 21st of March, the World Meteorological Organisation (WMO) released a comprehensive report summarising the available research evidence on the state of planet's climate up to the end of 2016 (World Meteorological Organisation 2017). As the accompanying press release noted, "The year 2016 made history, with a record global temperature, exceptionally low sea ice, and unabated sea level rise and ocean heat...2016 was the warmest on record—a remarkable 1.1 °C above the pre-industrial period, which is 0.06 °C above the previous record set in 2015... Carbon dioxide levels in the atmosphere reached the symbolic benchmark of 400 parts per millions in 2015—the latest year for which WMO global figures are available—and will not fall below that level for many generations to come because of the long-lasting nature of $CO_2$." These stark figures prompted David Carlson, the director of the WMO's climate research program to note that "We are seeing remarkable changes across the planet that are challenging the limits if our understanding of the climate system. We are now truly in uncharted territory" (Carrington 2017).

Worsening conditions in two of the main areas or concern, polar ice and sea temperatures, were confirmed by authoritative specialised studies published within days of the WMO report. Research from the National Snow and Ice Data Centre (2017) confirmed that the maximum extent of the Arctic ice sheet in September 2016 was the second lowest for any year in the 38 years for which there are satellite records, just above the lowest recorded level in 2012, with major impacts on extreme weather conditions across North America, Europe and Asia, including China, where independent research linked arctic sea ice loss to the trapped stagnant air fuelling extreme winter air pollution (Zou et al. 2017). This work points up the general relation between atmospheric pollution and the global warming caused by the prevailing organisation of market capitalism but a second study published in March detailed the number of premature deaths directly attributable to the international production and trade in consumer goods (Zhang et al. 2017). The authors estimate that 22% of the almost 3.5 million annual pollution related deaths globally are linked to exported goods and services. They focus on the fine particle matter pollution (PM2.5) that is estimated to account for 90% of the premature global deaths associated with air pollution. Analysing emissions from factories, transportation and energy generation in 13 regions, they identified two processes at work. Firstly, faced with competition to satisfy the entrenched demand for cheap consumer products in the affluent West, manufacturing centres in emerging economies are unlikely to fully enforce relevant environmental and health and safety regulations. Secondly, pollution generated in specific locations is never only a local issue since it is dispersed to other regions on global air currents. They calculated that 108,600 premature deaths in China can be linked to consumer demand in North America and Europe, including demand for smart phones and other digital devices, adding to the deaths attributable to the hazards associated with salvaging saleable materials from electronic waste at the other end of the production cycle.

New research released in March addressed another fundamental climate process, the accelerating ocean warming that contributes significantly to overall planetary temperatures, the likelihood of intense rainfall events and the bleaching of coral reefs (Cheng et al. 2017). Employing improved techniques for measuring ocean temperatures the authors demonstrated that over the last 60 years the oceans have been warming 13% faster than previously thought and that since 1990 warming has penetrated to depths below 700 metres. The world's most famous reef is the Great Barrier off the north-east coast of Australia. As a government commissioned report on the state of the Australian Environment published in March 2017 noted, record high water temperatures have caused widespread coral bleaching, habitat destruction and species mortality between 2011–2016 (Jackson et al. 2017: xii). In 2016, mass bleaching killed 22% of the coral on the Great Barrier Reef, with an aerial survey on 09 March 2017 confirming that a second mass bleaching event was underway caused by an underwater heat wave (The Great Barrier Reef Marine Park Authority 2017). The construction of the planned Carmichael coal mining complex, addressed by Benedetta Brevini and Terry Woronov in their chapter, will have a major adverse impact.

## CONVENIENT UNTRUTHS

The complex, comprising six open-cut pits and additional underground mines, would be the world's most extensive exploitation of untapped global coal reserves. The project is being proposed by The Adani Group, a diversified conglomerate headed by one of India's wealthiest and most influential men, Gautam Adani. It comes at a time when there is a growing consensus that halting all new extraction and 'keeping it in the ground', in the resonant slogan promoting *The Guardian's* campaign (described in Alan Rusbridger interview), is absolutely essential if the target for cutting emissions set at the Paris Summit have any chance of being realised. And as Michael E. Mann points out in his interview in this volume, this target only gets us part of the way to the reductions necessary to avoid catastrophe. The environmental destruction at the extraction site would be compounded by a proposed 400 kilometre rail link to the Queensland coast at Abbot Point, by the offshore processing facilities, and by the shipping required to transport the finished coal to its final destination in India.

The project has unreserved support from the Queensland and Federal governments. In a theatrical gesture in February the national treasurer, Scott Morrison, brandished a lump of coal during Question Time in the House of Representatives taunting opponents and declaring, 'This is coal. Don't be afraid, don't be scared.' He elaborated on his position in a radio interview a few days later stressing that the government "have no more fear of coal than we have a fear of wind, or solar, or wave energy, or pump-hydro, or whatever the option is" (Hunt 2017). This statement, which positions coal as simply another option for energy generation alongside renewables, is a prime example of the false equivalents at the centre of opposition to the overwhelming scientific evidence on the causes and consequences of climate crisis.

In March the dedicated climate change denier, James Delingpole, posted a commentary on the alt-right news site, *Breitbart*, attacking a *Washington Post* story reporting the scientific evidence of bleaching on the Great Barrier Reef (mentioned earlier) asserting that it was "just a #fake-news lie designed to promote the climate alarmist agenda" (Delingpole 2017). He claims to have been "out there personally to check" and found that it is not "in any kind of danger". He doesn't say when he visited or which part of the reef he went to, but has no hesitation in presenting his casual observations as equivalent to the weight of systematic scientific inquiry. He goes on to attribute the informed consensus on bleaching to a "green-left-liberal echo chamber" singling out "anyone who works for the ABC in Australia, the BBC, *The Guardian*" as leading promoters of his conception of fake news on the issue. This list is instructive. The ABC and BBC are both public service broadcasters and *The Guardian* is operated by a trust and has no proprietor or shareholders. It is precisely the distance from overt corporate influence these arrangements provide that opens space for critical inquiry into competing claims.

A House of Commons Committee report on science coverage in the British media published in March quoted the conclusion of an earlier study that had found that the "BBC content was generally of a high quality and exemplary in its precision and clarity" (House of Commons 2017: 3), but noted that the statutory obligation to observe due impartiality and provide balance between alternative views can create the impression that the contending positions command equal weight" (ibid.: 14). As they point out however, "the print and other media" are under no obligation to be balanced and "often have an agenda with inadequate place for opposing evidence" (ibid.: 3). An earlier inquiry into the coverage of climate change had found that this partisanship generated coverage that was inaccurate, 'inherently biased', and driven by a desire to overstate claims made by individuals, with no scientific training, who denied the evidence on climate crisis (ibid.: 13).

This refusal to accept the evidential weight behind the overwhelming scientific consensus on the causes of climate crisis has been accompanied by persistent attacks on attempts to interrogate the claims underpinning advocacy of continued reliance on fossil fuels. In late 2016, the ABC broadcast investigations into the Carmichael mine project including an analysis of the Adani conglomerate drawing attention to the Indian authorities' continuing investigations into alleged money laundering and tax fraud by Adani subsidiaries. A few days later, the Minister for Resources, Matt Canavan, interviewed on an ABC morning radio show, declared that he had "been very disappointed in the ABC's coverage" adding that "your reports have been nothing but fake news" (Landers 2016). In February 2017, the national government disregarded the short list for new members of the ABC's governing board drawn up by the independent nominating panel, and appointed Vanessa Guthrie, the first female chair of the Minerals Council, one of the most influential groups lobbying on behalf of the mining and resources industries.

Government antagonism to public service media has also intensified in the United States since Donald Trump's election as President. An analysis of the coverage of climate change on the main broadcast television networks, published in March 2017, showed a sharp decline. In 2015, the year that saw the warmest year on record and the signing of the Paris climate agreement, the three established networks (ABC, CBS, NBC) and Fox broadcast just under two and half hours of coverage (147 minutes) between them. In 2016, the year of the presidential election, that total had dropped to 40 minutes. The sole exception was the public service channel PBS, which carried 46 segments on climate change, ten more than the total for the established three commercial networks (Kalhoefer 2017). President Trump's preliminary 2018 budget proposal, issued on 16 March 2017, included a plan to eliminate completely the $445 million of federal funding for the Corporation for Public Broadcasting that supports both PBS and public radio.

This squeeze on independent media scrutiny was accompanied by a $900 million funding cut to the Office of Science that supports essential research and a major reduction in the scope of the Environmental Protection Agency (EPA), the main governmental body addressing climate related issues. If implemented the budget would see the Agency's funding slashed by 31%, its workforce reduced by a fifth, and 50 or so programs eliminated, including grants to support states and cities in combatting air pollution.

These moves follow a series of presidential orders removing restrains on the fossil fuel industries. Regulations barring surface mining companies from polluting waterways and compelling coal and energy companies to pay more in federal royalties have been set aside, federal officials are no longer required to consider environmental impacts when making decisions, and the EPA has been instructed to dismantle President Obama's Clean Power Plan. This signature intervention requiring states to reduce overall emissions and limiting carbon emission from power plants was announced in advance of the Paris climate summit as a concrete demonstration of the US's determination to take the lead in combating climate crisis. Four weeks before the summit, President Obama moved to bolster this claim by turning down TransCanada's proposal to build the Keystone XL pipeline transporting oil from the Albert tar sands over a thousand miles to Steele City in Nebraska where it would be transferred to another pipeline taking is down to refineries on the gulf coast. The proposed Keystone route would pass over an underground reservoir on the Great Plains, the main water source for the region's Native American communities. President Obama's intervention ended years of controversy and protest over possible pollution and other concerns. Announcing his decision he reiterated the claim that "America is now a global leader when it comes to taking serious action on climate change. Frankly, approving that project would have undercut that global leadership, and that is the biggest risk we face: not acting" (Jamieson 2017).

President Trump's calculated assault on Obama's measures to combat climate crisis has the enthusiastic support of Scott Pruitt, his appointee as head of the EPA. Pruitt has a long record of challenging environmental regulations. During his tenure as attorney general of Oklahoma, he mounted challenges to EPA provisions on 14 occasions, often in collaboration with fossil fuel company interests. Interviewed in his new position by CNBC at the beginning of March he dismissed the incontrovertible research evidence linking global warming to emissions of carbon dioxide reiterating the entirely erroneous claim that scientist could not agree on the impact, telling the interviewer:

> that measuring with precision human activity on the climate is something very challenging to do and there's tremendous disagreement about the degree of impact, so no, I would not agree that it's a primary contributor to the global warming that we see (Milman 2017).

This put him at odds with his own agency, whose web site clearly states in the section headed "Causes of Climate Change" that:

> Recent climate changes, however, cannot be explained by natural causes alone. Research indicates that natural causes do not explain most observed warming, especially warming since the mid-20th century. Rather, it is extremely likely that human activities have been the dominant cause of that warming (United States Environmental Protection Agency 2017).

It also puts him at odds with the 68% of Americans who, according to a Gallup poll published on 14 March 2017, now accept that global warming is caused by human activities and the 71% who agree that "most scientists believe global warming is occurring" (Saad 2017). These figures however still leave around a third of American citizens refusing to endorse the majority opinion.

One explanation for this is the deepening crisis of public communication, which is making it increasingly difficult to counter dismissals and misrepresentations of the scientific research supporting the urgency of taking comprehensive action to address climate crisis. Evidence and argument demonstrating the destructiveness of the fossil fuel industries is now routinely labelled as 'fake', not only by militant climate change deniers on alt-right web sites but, as we have seen here, by Government Ministers. In his landmark 2006 documentary, Al Gore characterized the mounting scientific evidence on climate change as 'An Inconvenient Truth'. The decade since then has seen research across the entire range of relevant scientific specialisms confirm the severity and extent of the impacts generated by carbon capitalism met by a proliferation of convenient untruths.

The fossil fuel industries have been able to rely on sympathetic press proprietors to support their interests for over a century. Added to which, as David McKnight and Mitchell Hobbs point out in their chapter detailing corporate efforts to derail the introduction of carbon trading schemes in Australia, 'dark money' supplied by investors in coal and mining has lent substantial financial support to a raft of think tanks and foundations publishing highly partial research and commentary masquerading as disinterested information. Against this, investigative journalism and specialist environmental and science reporting by public service broadcasters and broadsheet newspapers has attempted to expose false claims and provide independent assessments of the weight of available evidence. This fragile, and always unequal ecology, has been interrupted by the rise of social media, with far reaching consequences.

Recent years have seen a rapid expansion in the number of websites promoting climate change denial as part of a wider advocacy of market fundamentalist strategies aimed at defending business as usual and reducing state intervention to an absolute minimum. The best known, and arguably most influential, is *Breitbart News*, launched in 2007 and managed since 2012 by Steve Bannon, currently the Chief Strategist in the Trump administration, who is on record as casting the mainstream media as 'the opposition party', a position Trump shares. The reach and influence of these far-right sites is amplified by their appearance on the news and information feeds offered by the two most used social media sites, Google and Facebook, including Google's quick answers to users' questions, a facility which is also installed on the company's stand-alone domestic appliance, Google Home (Jeffries 2017). By eliminating the need to conduct a conventional search for relevant information this service replaces the option of evaluating competing sources with a single predetermined answer. The business model employed by both Google and Facebook is based on harvesting and analysing user data for resale to advertisers wanting to target their appeals more precisely. This promise has attracted a rapidly increasing share of advertising spending leaving many conventional printed newspapers in a precarious financial position. Taken together these shifts in the communications environment have major consequences for the mobilization of public support for fundamental measures to address climate crisis. Three interventions that would make a difference are currently under active discussion.

Firstly, as key contemporary gatekeepers employing algorithms to direct customized information and comment to users' home pages, the leading social media corporations must accept that they are publishers in their own right and not simply neutral technological 'platforms' for distributing content produced elsewhere (Bell and Owen 2017), and assume full responsibility postings on their sites, a requirement backed by fines for carrying material that circulates demonstrable untruths.

Secondly, investigative inquiry and specialist environmental reporting in the press needs to be supported by subsidies paid for out of the public purse rather than relying solely on donations and subscriptions from readers.

Thirdly, public service broadcasting's relative autonomy from state control and its entitlement to an income free from the need to accept advertising and sufficient to fund comprehensive investigation and analysis, needs to be vigorously defended and guaranteed. At the same time, public broadcasters should be supported in extending their public service remit to position themselves as central nodes in a digital network of public institutions offering access to comprehensive information, analysis and reasoned debate on key issues (see Murdock 2016). As the central argument of this book makes clear however, any proposal to develop new ways of organizing public information, participation and debate needs to ensure that it is not piggybacking on a communications infrastructure whose operation and use exacerbates climate crisis (see also Cubitt 2017: 168). This vision of a new digital commons is part of a much wider re-imagining of economic, social, and cultural relations.

## REASSERTING THE COMMONS

The recent appropriations of land and resources by fossil fuel companies are the latest instances in a process of commercial enclosure whose history stretches back to the origins of industrial capitalism and colonial adventurism. The foundational promotion of private property rights required the wholesale destruction and replacement of previous forms of economic and social organisation built around the idea of the commons.

At a practical level the commons fostered shared access to clusters of resources considered necessary to sustaining life and well-being; grazing rights, fishing rights, access to timber, plants and foodstuffs in forests and woodlands. Contrary to influential arguments supportive of marketisation, which have seen shared access leading to the progressive depletion of resources as individuals compete to maximize their personal advantages, commoning is anchored in a world view constructed around values of custodianship and sustainability rather than exploitation.

Indigenous peoples have often born the brunt of the dispossessions entailed in commercial logging, forest clearances, and coal and mining projects, but the ethical foundations of commoning have continued to provide a powerful imaginative counterweight to the force of capitalist enclosure. As Julian BraveNoiseCat, of the Salish peoples of what is now British Columbia noted in a piece published on 27 March 2017:

> Indigenous epistemologies were all but eliminated by colonization. British and American empires dispossessed indigenous people of their lands in the name of property and productivity. Despite this brutal and enduring history, indigenous people today stand on the frontlines of global movements fighting for a more just relationship between humanity and the land (BraveNoiseCat 2017).

As Naomi Klein points out in the interview included in this volume, it is the people who still live on the land and who are most directly affected by mining, pipelines and other fossil fuel projects who are leading the struggle against carbon capitalism and for alternative practices based on the ethos of the commons.

In 2016, the largest gathering of the Sioux nation since the meeting before the Battle of Little Big Horn convened on the Standing Rock reservation to protest plans to run the Dakota Access oil shale pipeline across native lands. They were acting not only to protect water supplies from possible contamination from leaks but to ensure that the resources they saw themselves holding in trust were handed on in good order. As the protest web site declared:

> In honor of our future generations we fight this pipeline to protect our water, our sacred places, and all living beings (Stand with Standing Rock 2017).

They won a landmark decision requiring the government to institute meaningful consultations on the protection of tribal lands and resources, only to see it overturned when President Trump signed an executive order to advance construction on the project, on the same day as giving the go-ahead to the Keystone XL pipeline.

These moves have been met with intensified opposition. Objectors lodged a suit in the federal court almost immediately after the Keystone decision was announced, arguing that the State Department had relied on an environmental impact assessment that was outdated and incomplete and therefore invalid. In a second suit a broad coalition of environmental groups alleged that failure to comply with the federal requirement that an environmental impact study be conducted prior to any decision being taken rendered Trump's lifting of the moratorium on new coal leases on federal land illegal. Indigenous rights were again at the heart of the case being made. As Jace Killsback, President of the North Cheyenne Tribe, whose lands in Montana would be most severely affected, made clear:

> The Nation is concerned that coal mining near the Northern Cheyenne Indian Reservation will impact our pristine air and water quality, will adversely affect our sacred cultural properties and traditional spiritual practices and ultimately destroy the traditional way of life that the Nation has fought to preserve for centuries (Frears and Ellperin 2017).

Indigenous rights are also central to current mobilizations against the Adani mine in Queensland detailed in Benedetta Brevini and Terry Woronov's chapter in this collection. The manifesto launching the broadly based Stop Adani Alliance in March 2017 includes 'respect for indigenous rights' as a key tenet alongside demands for the cancellation of the project, a ban on all new coal mines and an end to public subsidies for polluting projects including the $1 billion concessional loan the federal government has promised to help fund the rail link from the mine complex to the coast (The Stop the Adani Alliance 2017). The Galilee Blockade Campaign, which is planning peaceful civil disobedience around the mine site, has undertaken to "promote or be involved in direct action in the Galilee Basin [only] if requested by the Wangan and Jagalingou Traditional Owners family Council, the indigenous custodians of the land" (Galilee Blockade Campaign 2017).

Appeals to the core principles of the commons are no longer simply rhetorical. They can now cite a powerful legal precedent in support. In a landmark judgement in mid-March the New Zealand House of Representatives passed the third and final reading of a bill granting the country's third longest river, the Whanganui, its own legal identity with the corresponding rights of a legal person. The decision marked the final settlement of dispute dating back to the 1870s based on the local indigenous people's demand that the state recognize its special relation to the river. As the minister in charge of negotiations noted, "I know the initial inclination of some people is that it's pretty strange to give a natural resource a legal personality. But it's no stranger than family trusts, or companies" (Farand 2017). The Whanganui settlement is already providing a potent point of reference. A week after the New Zealand parliamentary vote, the high court in the Himalayan state of Uttarakhand cited it as a precedent in support of their decision to grant the status of 'living human entities' to the Ganges and Yamura rivers.

The fact that all three rivers have special spiritual significance for the local peoples will almost certainly be used by proponents of business as usual to limit the scope of future claims and present them as hang-overs from a pre-modern past with little relevance to contemporary conditions. But the idea of the commons is not so easily circumscribed. At its heart is a fundamental opposition between the continued corporate exploitation of resources and peoples in the pursuit of profit maximization and alternatives based on custodianship, collaborative production, and the equitable and just distribution of both rewards and costs. Working out how these alternatives might be organized is the most urgent question now facing anyone concerned to dismantle the continuing destructive impacts of carbon capitalism.

Pursuing this project arguably requires a shift in perspective. Environmental campaigning has tended to underline the urgency of taking action by promoting stark images of disaster, typified by the iconic photograph of a polar bear mentioned in the introduction to this volume, clinging to a small sliver of ice, staring out into a dark sea where until recently the ice sheet stretched to the far horizon. Recent Gallup polling suggests that these images, and the continual stream of climate disaster fiction, are escalating public anxiety with 45% of Americans polled in early 2017 saying that they now worried 'a great deal' about global warming, a significant increase on the 25% who gave the same answer in 2011 (Saad 2017). But, as Jodi Dean reminds us in her chapter, apocalyptic visions can have the effect of inducing resignation and even a perverse pleasure in witnessing the onwards march of destruction. While demonstrations of the severity of climate change impacts remain a necessary reminder of the scale of the problems facing us, campaigning also needs to emphasize that another world is indeed possible by celebrating victories in the struggle against fossil fuels and the success of practical alternatives: the local communities now generating sufficient power from renewable resources to be self-sufficient; the successful co-operatives reorganizing production, and the movements to repair and reuse rather than discard electronic consumer durables. In a communications environment increasing organized around resonant images and narratives we need to develop more effective cultural resources for hope that translate expert argument into memorable images and engaging stories using every available popular expressive form, from wall art and cartoons to animations and computer games.

Contests over the organization of public knowledge and argument around climate crisis remain central but as the chapters collected here make clear, a full analysis of the role of communication in sustaining carbon capitalism and possible ways of moving beyond it must also interrogate the impact of the material infrastructures, organizational complexes and assemblies of media devices, which support and direct communication. As Vincent Mosco reminds us in his chapter, the arrival of the Internet of Things and the increasing diversification of the leading social media companies into cloud computing, artificial intelligence and robotics, is installing these corporations as central forces in shaping the direction of advanced capitalism.

Current developments significantly increase the contribution that communication companies currently make to climate crisis, detailed by Richard Maxwell and Toby Miller in their chapter here. The new digital complexes place escalating demands on energy, water and resources in their production, transportation and use, reinforce the culture of hyper consumerism detailed by Justin Lewis in his chapter in this collection, and add to the accumulating amounts of waste and pollution already generated by accelerating rates of digital obsolescence and disposal (see Gabrys 2013; Grossman 2007). Faced with this new reality, we need more than ever to ask who should own and control the essential technological and organizational building blocks of communication, for what purposes, and with what consequences for collective well being?

# REFERENCES

Abbott, T. (2013, August 25). Tony Abbott's campaign launch speech: Full transcript. *Sydney Morning Herald*. Retrieved from http://www.smh.com.au/federal-politics/federal-election-2013/tony-abbotts-campaign-launch-speech-full-transcript-20130825-2sjhc.html.

ABCnet. (2016a). Adani faces more legal action as traditional owners vow to halt Carmichael coal mine. Retrieved from http://www.abc.net.au/news/2016-12-07/further-legal-action-planned-against-carmichael-coal-mine/8100326.

ABCnet. (2016b). Adani's $2 billion Carmichael rail line in Queensland closer to federal loan. Retrieved from http://www.abc.net.au/news/2016-12-03/adani-carmichael-rail-line-closer-to-federalloan/8089790.

Adeland. (2011, September 14). A call to action: Non-violent civil disobedience against the tar sands. Rabble.ca. http://rabble.ca/news/2011/09/call-action-non-violent-civil-disobedience-against-tar-sands.

Amos, C., & Swann, T. (2015). *Carmichael in context*. Australian Institute: Quantifying Australia's threat to climate action. 2015.

Arctic Council. (2016). *Arctic Resilience Report*. In M. Carson & G. Peterson (Eds.), Stockholm Environment Institute and Stockholm Resilience Centre. Stockholm. http://www.artctic-council.org/arr.

Artists and Climate Change. https://artistsandclimatechange.com/about/.

Asian Correspondent. (2017, March 16). New Zealand transforms Whanganui River into a legal person.

Askanius, T., & Uldam, J. (2011). Online social media for radical politics: Climate change activism on youtube. *International Journal Of Electronic Governance4*, (1–2).

Atton, C., & Hamilton, J. (2008). *Alternative journalism*. London: Sage.

Australian Associated Press. (2009). Coal industry says jobs, mines to be lost under ETS.

Australian Coal Association. (2011). Why should thousands of Australians lose their jobs if it makes no difference to global greenhouse emissions? [Advertisement] *The Sydney Morning Herald*, 5.

Australia Institute. (2016). *Coal wars: Facts check*, 107, 848. Retrieved from http://www.tai.org.au/sites/defualt/files/P303%20Coal%20hard%20facts_0. pdf#page=1&zoom=auto.

Australian Trade and Industry Alliance. (2011). Business is already tight and there's bad news on the world economy. It's the worst possible time for a carbon tax. [Advertisement]. *The Australian*, 7.

Ball Windspeaker, D. P. (2014). Neil Young raises $500k for anti-oil sands fight. *Windspeaker*. http://www.ammsa.com/publications/windspeaker/neil-young-raises-500k-anti-oil-sands-fight.

Banning, M. E. (2009). When poststructural theory and contemporary politics collide—The vexed case of global warming. *Communication and Critical/Cultural Studies*, 6(3), 285–304.

Barnes, B. (2010, March 4). The Na'vi Take on the tar sands. *New York Times*. http://carpetbagger.blogs.nytimes.com/2010/03/04/the-navi-take-on-the-tar-sands/.

Bates, J., & Goodale, P. (2017). Making data flow for the climate risk markets. *Television and New Media* (forthcoming).

Beers, D. (2006). The public sphere and online, independent journalism. *Canadian Journal of Education*, 29, 109–130. Retrieved from: http://www.csse.ca/CJE/Articles/CJE29-1.htm.

Bell, E., & Owen, T. (2017). *The Platform Press: How Silicon Valley reengineered journalism*. New York: Columbia Journalism School, Tow Centre for Digital Journalism.

Benson, P., & Kirsch, S. (2010). Corporate oxymorons. *Dialectical Anthropology*, 34(1), 45–48.

Bernays, E. L. (1955). The theory and practice of public relations: A resume. In E. L. Bernays (Ed.), *The engineering of consent* (pp. 3–25). Norman: University of Oklahoma Press.

Blainey, G. (2001). *This land is all horizons: Australian fears and visions*. London: Virago Press.

Blewett, R. (2012). Shaping a nation: A geology of Australia. Geoscience Australia and ANU E-Press.

Boykoff, M. T. (2008). The cultural politics of climate change discourse in UK tabloids. *Political Geography*, 27(5), 549–569.

Bonneuil, C., & Fressoz, J.-B. (2016). *The shock of the Anthropocene: The earth, history and us*. London: Verso.

Bourassa, E., Amend, E., & Secko, D. M. (2013). A thematic review and synthesis of best practices in environment journalism. *Journal of Professional Communication*,3(1), 39–65.

Boykoff, M., & Boykoff, J. (2004). Balance as bias: Global warming and the US prestige press. *Global Environmental Change*, 14, 125–136.

Brack, D. (2017). *The impacts of the demand for woody biomass for power and heat on climate and forests*. London: Chatham House. The Royal Institute of International Affairs.

Braman, S. (2006). *Change of state: Information, policy, and power*. Cambridge, Mass.: MIT Press.

BraveNoiseCat, J. (2017, March 27). The western idea of private property is flawed. Indigenous peoples have it right. *The Guardian*. https://www.theguardian.com/commentisfree/2017/mar/27/western-idea-private-property-flawed-indigenous-peoples-have-it-right.

Breusch, J. (2009). Rudd's climate compromise. *The Australian Financial Review*, 1.

Brevini, B. (2016). The value of environmental communication research. *International Communication Gazette, 78*(7).

Bridgwater, A. (2016). Nutanix CEO: Cloud computing will become a 'regular' utility, but it's complicated. *Forbes*. http://www.forbes.com/sites/adrianbridgwater/2016/11/09/nutanix-ceo-cloud-computing-will-become-a-regular-utility-but-its-complicated/#6d8ebe1c4cd0. Accessed 9 Nov 2016.

Brockington, D. (2009). *Celebrity and the environment: Fame, wealth and power in conservation*. London: ZED Books.

Bündnis 90/Die Grünen [Alliance 90/The Greens]. (nd). The arts and culture need freedom. http://www.gruene-bundestag.de/service-navigation/english/culture_ID_377806.html. Accessed 19 Jan 2017.

Burgoon, M., Pfau, M., & Birk, T. (1995). An inoculation theory explanation for the effects of corporate issue/advocacy advertising campaigns. *Communication Research, 22*, 485–505.

Burtt, N. (2015). French laws aim to outlaw planned obsolescence. *AIBM News*, 8 May, 2015. http://www.theiabm.org/news/news_detail.french-laws-aim-to-outlaw-planned-obsolescence.html.

Calcutt, A., & Hammond, P. (2011). *Journalism studies: A critical introduction*. London and New York: Routledge.

Cammaerts, B. (2012). Protest logics and the mediation opportunity structure. *European Journal of Communication, 27*(2), 117–134.

Campbell, B. (2015). Senate hearing, Queensland Government Administration related to Commonwealth Government Affairs Parliament of Queensland, 4th of February 2015.

Canadian Press. (2010, April 20). James Cameron slams tar sands as 'black eye' on Canada. *CTV News*. http://www.ctvnews.ca/james-cameron-slams-tar-sands-as-black-eye-on-canada-1.504039.

Canada Press. (2011, September 11). Fort McMurray radio station bans Neil Young songs for day, after singer blasts oilsands. *CTV News*. http://www.ctvnews.ca/entertainment/fort-mcmurray-radio-station-bans-neil-young-songs-for-day-after-singer-blasts-oilsands-1.1449801.

Canada Press. (2014, August 24). Leonardo DiCaprio nominates Stephen Harper to do ice bucket challenge. *CBC News.* http://www.cbc.ca/news/politics/leonardo-dicaprio-nominates-stephen-harper-to-do-ice-bucket-challenge-1.2746908.

CAPP. (2017). "2017 Crude oil forecast, markets and transportation." Webpage. *Canadian Association of Petroleum Producers (CAPP).* Retrieved June 28, 2017, from http://www.capp.ca/publications-and-statistics/crude-oil-forecast.

Castells, M. (2009). *Communication power.* Oxford: Oxford University Press.

Carpenter, de Vries, W., de Wit, C.A., Folke, C., Gerten, D., Heinke, J., Mace, G. M., Persson, L. M., Ramanathan, V., Reyers, B., Sörlin, S. (2015). Planetary boundaries: Guiding human development on a changing planet. *Science, 15* (online, np.). doi:10.1126/science.1259855

Carrington, D. (2017, March 21). Record-breaking climate change pushes world into 'uncharted territory. *The Guardian.* https://www.theguardian.com/environment/2017/mar/21/record-breaking-climate-change-world-uncharted-territory.

Carrington, D., & Mommers, J. (2017, February 28). We must act on climate change, Shell declared-back in 1991. *The Guardian, 3.*

Carvalho, A. (2007). Ideological cultures and media discourses on scientific knowledge: Re-reading news on climate change. *Public Understanding of Science, 16*(2), 223–243.

CBC National. (2010, March 6). Avatar seen as eco parable. *The National.* http://www.cbc.ca/news/thenational/avatar-seen-as-eco-parable-1.1795775.

CBC News. (2013, March 6). The chinook: Calgary's reprieve from the grasp of winter. *CBC Calgary.* http://www.cbc.ca/calgary/interactives/chinook/.

CBC News. (2014, January 16). Neil Young's anti-oilsands comments draw fire from industry. *CBC News.* http://www.cbc.ca/news/canada/calgary/neil-young-s-anti-oilsands-comments-draw-fire-from-industry-1.2499176.

Cellan-Jones, R. (2015). Office puts chips under staff's skin. *BBC News.* http://www.bbc.com/news/technology-31042477. Accessed 29 Jan 2016.

Centre for Energy-Efficient Telecommunications. (2013). *The power of wireless cloud* (White paper). http://www.ceet.unimelb.edu.au/pdfs/ceet_white_paper_wireless_cloud.pdf. Accessed 19 Jan 2017.

Cheng, L., Trenberth, K., et al. (2017, March 25). Improved estimates of ocean heat content from 1960 to 2015. *Science Advances, 3.* http://advances.sciencemag.org/content/3/3/e1601545.

Christensen, G. (2015). Speech at House of Parliament, Hansard, 19th August 2015.

Christensen, G. (2016). Speech at House of Parliament, Hansard, 8th of February 2016.

Christians, C., Glasser, T., McQuail, D., Nordenstreng, K., & White, R. A. (2009). *Normative theories of the media: Journalism in democratic societies.* Urbana and Chicago: University of Illinois Press.

Chubb, P. (2014). *Power failure: The inside story of climate politics under Rudd and Gillard.* Collingwood: Black Inc.

Churchman, C. W. (1967). Wicked problems. *Management Science, 14,* B141–B142.

Cleary, P. (2011). *Too much luck: The mining boom and Australia's future.* Melbourne: Black Inc.

Cleary, P. (2015). Coal crash: What it means for Australia. *The Monthly,* 23–31.

Consumers for Safe Phones. (2011, November 27). FCC's cell phone testing dummy is larger than 97% of all cell phone users. http://consumers 4safephones.com/fccs-cell-phone-testing-dummy-is-larger-than-97-of-all-cell-phone-users/. Accessed 19 Jan 2017

Cook, B. I., Anchukaitis, K. J., et al. (2017). Spatiotemporal drought variability in the Mediterranean over the last 900 years. *Journal of Geophysical Research, 122* (5), 2060–2074.

Coombs, W. T. (1998). An analytic framework for crisis situations: Better responses from a better understanding of the situation. *Journal of Public Relations Research, 10,* 177–191.

Coombs, W. T. (1999a). Information and compassion in crisis responses: A test of their effects. *Journal of Public Relations, 11,* 125–142.

Coombs, W. T. (1999b). *Ongoing crisis communication: Planning, managing, and responding.* Thousand Oaks CA: Sage.

Cooper, M. (2010). Turbulent worlds: Financial markets and environmental crisis. *Theory, Culture & Society, 27*(2–3), 167–190.

Coorey, P. (2011). Abbott will eat his words on carbon, PM says. *The Sydney Morning Herald,* 7. Costanza, R. et al. (2014). Development: Time to leave GDP behind. *Nature.* http://www.nature.com/news/development-time-to-leave-gdp-behind-1.14499.

Corner, J., & Pels, D. (Eds.). (2003). *Media and the restyling of politics: Consumerism, celebrity and cynicism.* Thousand Oaks, CA: Sage.

Cottle, S. (2009). Series editor's preface: Global crises and the media. In T. Boyce & J. Lewis (Eds.), *Climate Change and Media* (pp. xiii-xvii). New York: Peter Lang.

Couldry, N. (2013). *Media, society, world: Social theory and digital media practice.* London: Polity.

Council of Canadians. (2011, September 27). UPDATE: How well did mainstream media cover our tar sands protest? *Council of Canadians.* http://canadians.org/fr/node/7661.

Council of Canadians. (2015). Tar sands. Retrieved June 20, 2016, from http://canadians.org/tarsands.

Cox, R. J. (2007). Nature's 'crisis disciplines': Does environmental communication have an ethical duty?, *Environmental Communication: A Journal of Nature and Culture, 1*(1), 5–20. Available online at http://www.informaworld.com.proxy. lib.sfu.ca/smpp/section?content=a778982174&fulltext=713240928. Accessed on 1 March 2013.

Crowe, D. (2011). Labor must reply to scares. *The Australian Financial Review*, 13.

Crutzen, P. J. (2002, January 3). Geology of mankind. *Nature, 415*, 23.

Crutzen, P. J., & Steffen, W. (2003). How long have we been in the Anthropocene era? An editorial comment. *Climatic Change, 61*, 251–257.

CTV News Staff. (2010, September 29). CTV—Oilsands can be a curse. http:// www.ctvnews.ca/oilsands-can-be-a-curse-to-canada-cameron-says-1.557924.

Cubitt, S. (2017). *Finite media: Environmental impacts of digital technologies.* Durham: Duke University Press.

Curran, J., Fenton, N., & Freedman, D. (2012). *Misunderstanding the internet.* London: Routledge.

Daly, M., Gifford, L., Luedecke, G., McAllister, L., Nacu-Schmidt, A., Andrews, K., et al. (2015). *World newspaper coverage of climate change or global warming, 2004–2015.* Center for Science and Technology Policy Research, Cooperative Institute for Research in Environmental Sciences, University of Colorado, Web. http://sciencepolicy.colorado.edu/media_coverage.

Davenport, T. (2014). Predictive analytics: A primer. *Harvard Business Review.* https://hbr.org/2014/09/a-predictive-analytics-primer.

Davoudi, S. (2000). Planning for waste management: Changing discourses and institutional relationships. *Progress in Planning, 53*, 165–216.

Denniss, R., Adams, P., Campbell, R., & Grudnoff, M. (2016). Never gonna dig you up! Modelling the economic impacts of a moratorium on new coal mines. Australia Institute, August 2016.

Delicath, J. W., & Delcua, K. M. (2003). Image events, the public sphere, and argumentative practice: The case of radical environmental groups. *Argumentation, 17*, 315–333.

Delingpole, J. (2017, March 20). Great Barrier Reef Still Not Dying, Whatever Washington Post Says. *Breitbart.* http://www.breitbart.com/big-government/ 2017/03/20/delingpole-great-barrier-reef-still-not-dying-whatever- washington-post-says/.

Della Porta, D., & Diani, M. (2006). *Social movements: An introduction.* Oxford and Victoria: Blackwell Publishing.

DeLuca, K. M. (1999). *Image politics: The new rhetoric of environmental activism.* London: Guilford Press.

Department for Business Enterprise and Regulatory Reform. (2008). Supporting innovation in services. London: HMSO. http://www.bis.gov.uk/files/ file47440.pdf. Accessed 7 Oct 2011.

Devra, D. (2013). War-gaming cell phone science protects neither brains nor private parts. *The huffington post*.http://www.huffingtonpost.com/devra-davis-phd/cell-phones-brain-cancer_b_3232534.html.

Dignan, L. (2011a). Cloud computing's real creative destruction may be the it workforce. *ZDNet*. http://www.zdnet.com/article/cloud-computings-real-creative-destruction-may-be-the-it-workforce/. Accessed 24 Oct 2011.

Dignan, L. (2011b). Analytics in 40 years: Machines will kick human managers to the curb. *ZDNet*. http://www.zdnet.com/article/analytics-in-40-years-machines-will-kick-human-managers-to-the-curb/. Accessed 18 Oct 2011.

Dischel, R. (2002). Introduction to the weather market: Dawn to mid-morning. In Robert Dischel (Ed.), *Climate risk and the weather market: Financial risk management with weather hedges* (pp. 3–24). London: Risk Books.

Doyle, J. (2011). *Mediating climate change*. Chicago: Ashgate Publishing, Ltd.

Dutton, J. A. (2002). The weather in weather risk. In R. Dischel (Ed.), *Climate risk and the weather market: Financial risk management with weather hedges* (pp. 185–214). London: Risk Books.

Dyer, S., Grant, J., & Angen, E. (2013). *Forecasting impacts of oilsands expansion*. Report. Pembina Institute. Retrieved June 16, 2016 from https://www.pembina.org/pub/2455.

Edwards, P. N. (2013). *A vast machine: Computer models, climate data, and the politics of global warming*. Cambridge, Mass.: MIT Press.

Eghbal, M. A., Eftekhari, A., Ahmadian, E., Azarmi, Y., & Alireza, P. (2016). A review of biological and pharmacological actions of melatonin: Oxidant and prooxidant properties. *Journal of Pharmaceutical Reports, 1*, 106–115.

Elango, N., Kasi, V., Vembhu, B., & Poornima, J. G. (2013). Chronic exposure to emissions from photocopiers in copy shops causes oxidative stress and systematic inflammation among photocopier operators in India. *Environmental Health, 12*, 78–90. Published online 2013 September 11. https://www.ncbi.nlm.nih.gov/pmc/articles/PMC3849716/. Accessed 19 Jan 2017.

Entman, R. M. (2004). *Projections of power: Framing news, public opinion, and US foreign policy*. Chicago: University of Chicago Press.

European Commission. (nd). Supporting cultural and creative industries. http://ec.europa.eu/culture/policy/cultural-creative-industries/index_en.htm. Accessed 19 Jan 2017.

European Commission. (2016). Waste electrical & electronic equipment (WEEE). http://ec.europa.eu/environment/waste/weee/index_en.htm. Accessed 19 Jan 2017.

European Environment Agency. (2017). *Climate change, impacts and vulnerability in Europe 2016: An indicator-based report*. EEA Report No 1, Luxembourg. Publications Office of the European Union.

Evans, G. (2015). Cornerstone export industry's outlook far from black. *Australian Financial Review*, 6.

Ewen, S. (2001). *Captains of consciousness: Advertising and the social roots of consumer culture* (25th ed.). New York: Basic.

Farand, C. (2017, March 15). New Zealand river becomes first in the world to be given legal status of a person. *The Independent.* http://www.independent.co.uk/news/world/australasia/whanganui-river-new-zealand-legal-status-person-north-island-worlds-first-maori-people-bill-iwi-a7630596.html.

Fawcett, L. (2002). Why peace journalism isn't news. *Journalism Studies, 3*(2), 213–223.

Featherstone, M. (2007). *Consumer culture and postmodernism.* London: Sage.

Fekete, J. (2016, January 29). "Oilsands' share of GHG emission to double by 2030". *Ottawa Citizen.* Retrieved June 20, 2016 from http://ottawacitizen.com/news/politics/oilsands-share-of-national-emissions-set-to-doublebetween-2010-and-2030.

Fessoz, J.-B. (2007). Beck back in the 19th century: Towards a geneology of risk society. *History and Technology, 23*(4), 333–350.

Forbes. (2017). The world's billionaires. https://www.forbes.com/billionaires/list/#version:realtime. Accessed 8 March 2017.

Forde, S. (2011). *Challenging the news: The journalism of alternative and community media.* New York: Palgrave Macmillan.

Forde, S. (2017). Chapter 7: Alternative approaches to environment coverage in the digital era: The Guardian's 'Keep it in the Ground' campaign. In R. A. Hackett, S. Forde, S. Gunster & K. Foxwell-Norton (Eds.), *Journalism for Climate Crisis: Public engagement, media alternatives* (pp. 167–189). London: Routledge.

Foxwell-Norton, K. (2017). Chapter 6: Australian independent news media and climate change reporting: The case of COP21. In R. A. Hackett, S. Forde, S. Gunster & K. Foxwell-Norton, *Journalism for Climate Crisis: Public engagement, media alternatives* (pp. 144–166). London: Routledge.

Frears, D., & Ellperin, J. (2017, March 30). The standoff between Trump and green groups just boiled into war. *The Washington Post.* https://www.washingtonpost.com/news/energy-environment/wp/2017/03/30/the-standoff-between-trump-and-green-groups-just-boiled-into-war/?tid=a_inl&utm_term=.fceec787d3b7.

Frey, C. B., & Osborne, M. A. (2013). *The future of employment: How susceptible are jobs to computerisation?* Oxford University. http://www.oxfordmartin.ox.ac.uk/publications/view/1314.

Friedland, L. A., & Nichols, S. (2002) *Measuring civic journalism's progress: A report across a decade of activity.* Washington, DC: Pew Center for Civic Journalism. Accessed at http://www.pccj.org/doingcj/research/measuringcj.pdf.

Frydenberg, J. (2015). Minister Frydenberg speech "Questions Without Notice - Resource Projects in the Electorate of Dawson", November 30th 2015.

Gabrys, J. (2007). Media in the dump. In J. Knechtel (Ed.), *Alphabet city: Trash* (pp. 156–165). Cambridge: MIT Press.

Gabrys, J. (2013). *Digital rubbish: A natural history of electronics*. Ann Arbor: The University of Michigan Press.

Gaffney, O., & Steffen, W. (2017). The Anthropocene equation. *The Anthropocene Review*. http://journals.sagepub.com/doi/full/10.1177/2053019616688022.

Gagliordi, N. (2015). Only 40 percent of the global population has ever connected to the internet: Report. *ZDNet*. http://www.zdnet.com/article/only-40-percent-of-the-global-population-ever-connected-to-the-Internet-report/. Accessed 25 Feb 2015.

Galilee Blockade Campaign. (2017). *Galilee Blockade Campaign: Together we will win*. galileeblockade.net/traditional owners.

Gallegos, D., & Narimatsu, J. (2015). ICT at COP21: Enormous potential to mitigate emissions. *The World Bank*. http://www.worldbank.org/en/topic/ict/brief/connections-note-30.

Gandy, O. H. (1993). *The panoptic sort: A political economy of personal information*. Boulder: Westview Press.

Giddens, A. (2006). *The politics of climate change*. Cambridge: Cambridge University Press.

Gilbert, C. J. (2016). On truth and lies in an affective sense. In J. Hannan (Ed.), *Truth in the public sphere* (pp. 93–114).

Gillespie, T. et al. (2014). *Media technologies: Essays on communication, materality, and society*. Cambidge, Mass: The MIT Press.

Giroux, H. A., & Bhattacharya, D. (2016). Anti-politics and the scourge of authoritarianism. *Social Identities*, 1–15

Gitlin, T. (1980). *The whole world is watching: Mass media in the making and unmaking of the new left*. Berkeley: University of California Press.

Government of Alberta. (2016). "Facts and statistics". Retrieved June 20, 2016, from http://www.energy.alberta.ca/oilsands/791.asp.

Goel, V., & Perlroth, N. (2016). Hacked yahoo data is for sale on dark web. *The New York Times*. http://nyti.ms/2gSokng. Accessed 16 Dec 2016.

Goldenberg, S. (2015). Edelman ends work with coal producers and climate change deniers. *The Guardian*. https://www.theguardian.com/environment/2015/sep/15/edelman-ends-work-with-coal-and-climate-change-deniers.

Goldstein, J. (2006). The remains of the everyday: One hundred years of recycling in Beijing. In M. Y. Dong & J. Goldstein (Eds.), *Everyday Modernity in China*. Seattle: University of Washington Press.

GoodElectronics. http://goodelectronics.org/. Accessed 19 Jan 2017.

Gordon, R. J. (2016). *The rise and fall of american growth*. Princeton, NJ: Princeton University Press.

Gower, K. (2008). US corporate public relations in the progressive era. *Journal of Communication Management, 12*(4), 305–318.

Greenberg, J., Knight, G., & Westersund, E. (2011). Spinning climate change: Corporate and NGO public relations strategies in Canada and the United States. *International Communication Gazette, 73*(1–2), 65–82.

Greengard, S. (2015). *The internet of things.* Cambridge, MA: MIT.

Green Party of Aotearoa New Zealand. (2012). Arts, culture and heritage policy summary. https://home.greens.org.nz/policysummary/arts-culture-and-heritage-policy-summary. Accessed 19 Jan 2017; UK Green Party. (2014). Culture, media, and sports. https://policy.greenparty.org.uk/culture,-media-and-sports.html. Accessed 19 Jan 2017.

Green Party of California. (nd). Arts and culture. http://www.cagreens.org/platform/arts-and-culture. Accessed 19 Jan 2017.

Green Party USA. (nd). Platform of the Greens/Green Party USA. https://www.greenparty.org/Platform.php. Accessed 19 Jan 2016.

Greenpeace. (nd). Green IT at Greenpeace. http://www.greenpeace.org/international/en/campaigns/climate-change/cool-it/Green-IT-at-Greenpeace/. Accessed 19 Jan 2017.

Greenpeace Canada. (2008a, July 23). Greenpeace activists interrupt Syncrude tar sands operation. *Greenpeace Canada.* http://www.greenpeace.org/canada/en/recent/greenpeace-activists-interrupt/.

Greenpeace Canada. (2008b). Greenpeace activists at Ed Stelmach's fundraising dinner. https://www.youtube.com/watch?v=c3K2QF_-zuI. Uploaded 24 April 2008.

Greenpeace Canada. (2008c—if quoted?). http://newsletter.greenpeace.ca/alerts/2008-04-24-tar-sands/.

Greenpeace Canada. (2015). Tar sands. Retrieved June 20, 2016, from Greenpeace, http://www.greenpeace.org/canada/en/campaigns/Energy/tarsands/.

Greenpeace International. (2012). *How green is your cloud?* http://www.greenpeace.org/international/Global/international/publications/climate/2012/iCoal/HowCleanisYourCloud.pdf. Accessed 19 Jan 2017.

Grigg, A. (2010a). Slick campaign presses the right buttons. *Australian Financial Review,* 24 June, 8.

Grigg, A. (2010b). Miners are her first big challenge. *Australian Financial Review,* 25 June, 1, 30.

Grigg, A. (2010c). $10 m splurge shows way for unhappy industries. *Australian Financial Review,* 3 July, 9.

Grossman, E. (2007). *High tech trash: Digital devices, hidden toxics, and human health.* Washington, DC: Shearwater/Island Press.

Grossman, Elizabeth. (2009). *Chasing molecules: Poisonous products, human health, and the promise of green chemistry.* Washington, DC: Shearwater/Island Press.

Grossman, E. (2016). The body burden—Toxics, stresses and biophysical health. In R. Maxwell (Ed.), *The Routledge companion to labor and media* (pp. 65–77). New York: Routledge.

Guardian. (2016). *The great barrier reef: A catastrophe laid bare.* Retrieved from https://www.theguardian.com/environment/2016/jun/07/the-great-barrier-reef-a-catastrophe-laid-bare.

Gunster, S. (2011). Covering copenhagen: Climate politics in B.C. media. *Canadian Journal of Communication, 36*(3), 477–502.

Gunster, S. (2012). Radical optimism: Expanding visions of climate politics in alternative media. In A. Carvalho & T. R. Peterson (Eds.), *Climate change politics: Communication and public engagement* (pp. 239–267). Amherst, N.Y.: Cambria Press.

Gunster, S. (2017a). Chapter 3: Engaging climate communication: Audiences, frames, values and norms. In R. A. Hackett, S. Forde, S. Gunster & K. Foxwell-Norton (Eds.), Journalism for Climate Crisis: Public engagement, media alternatives (pp. 49–76). London: Routledge.

Gunster, S. (2017b). Chapter 5: Contesting conflict? Efficacy, advocacy and alternative media in British Columbia. In R. A. Hackett, S. Forde, S. Gunster & K. Foxwell-Norton (Eds.), Journalism for Climate Crisis: Public engagement, media alternatives, London: Routledge, pp. 120–143. Accessed Jan. 19, 2017.

Gusterson, H. (2015). *Drones: Remote control warfare.* Cambridge, MA: MIT Press.

Hackett, R. A. (2006). Is peace journalism possible? In *Conflict and communication online* (Fall). Also In Shinar, D. & Kempf, W. (Eds.). (2007). *Peace journalism: The state of the art* (pp. 75–94). Berlin: Verlag Irena Regener.

Hackett, R. A. (2017a). Chapter 4: From frames to paradigms: Civic journalism, peace journalism and alternative media. In Hackett, R.A., Forde, S., Gunster, S., & Foxwell-Norton, K. (Eds.), *Journalism for climate crisis: Public engagement, media alternatives* (pp. 94–119). London: Routledge.

Hackett, R. A. (2017b). Conclusion: Media reform for climate action. In Hackett, R. A., Forde, S., Gunster, S. & Foxwell-Norton, K. (Eds.), *Journalism for climate crisis: Public engagement, media alternatives* (pp. 190–197). London: Routledge.

Hackett, R. A., Wylie, S., & Gurleyen, P. (2013). Enabling environments: Reflections on journalism and climate justice. *Ethical Space, 10*(2/3), 34–46.

Hackett, R. A., & Zhao, Y. (1998). *Sustaining democracy? Journalism and the politics of objectivity.* Toronto, Garamond [now University of Toronto Press].

Hackett, R. A., Wylie, S., & Gurleyen, P. (2013). Enabling environments: Reflections on journalism and climate justice. *Ethical Space, 10*(2/3), 34–46.

Hall, D. (2009). CPRS 'just crazy' coates. *Australian Mining,* 1.

Haluza-DeLay, R., Ferber, M., & Wiebe-Neufeld, T. (2013). Watching avatar from AvaTar Sands land. In B. Taylor (Ed.), *Avatar and nature spirituality* (pp. 123–140). Waterloo: Wilfrid Laurier University Press.

Hannan, J. (2016). Truth as first casualty in American politics. Introduction in Hannan, J. (Ed.), *Truth in the public sphere* (pp. XX11–XXX2). New York: Lexington.

Hansard. (2015). *Senator Brandis reponse to Senator Waters.* Retrieved from http://parlinfo.aph.gov.au/parlInfo/search/display/display.w3p;query=Id% 3A%22chamber%2Fhansards%2Fc3d90ba0-0cf0-40f5-b619-9d3ed6ad10bf% 2F0099%22.

Harrison, K. (2011). *Strategic public relations.* Melbourne: Palgrave MacMillian.

Harsin, J. (2015). Regimes of posttruth, postpolitics, and attention economies. *Communication, Culture and Critique, 8*(2), 327–333.

Harvey, D. (2011). *The enigma of capital and the crises of capitalism.* Oxford: Oxford University Press.

Hay, C., & Payne, T. (2013). *The great uncertainty.* Sheffield Political Economy Research Institute Paper 5. Sheffield: University of Sheffield. Available at: http://speri.dept.shef.ac.uk/wp-content/uploads/2013/01/SPERI-Paper-No.5-The-Great-Uncertainty-389KB.pdf.

Heath, R. L., & Waymer, D. (2011). Corporate issues management and political public relations. In J. Strömbäck & S. Kiousis (Eds.), *Political public relations: Principles and applications* (pp. 138–156). New York: Routledge.

HM Government. 2011. *Further detail on open data measures in the autumn statement 2011, London.* Available at: http://www.cabinetoffice.gov.uk/sites/ default/files/resources/Further_detail_on_Open_Data_measures_in_the_ Autumn_Statement_2011.pdf.

Holmes, D. (2016). The fossil-fuelled political economy of Australian elections. *The Conversation.* https://theconversation.com/the-fossil-fuelled-political-economy-of-australian-elections-61394.

Hope, B., & Vaishampayan, S. (2015, July 8). Glitch Freezes NYSE trading for hours. *Wall Street Journal.* http://www.wsj.com/articles/trading-halted-on-new-york-stock-exchange-1436372190.

House of Commons Science and Technology Committee. (2017, March 15). *Science Communication and Engagement: Eleventh Report of Sessison 2016–17.* London, House of Commons. HC162.

Howlett, D. (2014). Exclusive: Computer economics study—Cloud saves 15 percent. *Diginomica.*http://diginomica.com/2014/02/13/exclusive-computer-economics-study-cloud-saves/. Accessed 13 Feb 2014.

Hunt, E. (2017, February 13). Scott Morrison and Ray Hadley laugh about coal prop: 'Great stunt'. *The Guardian.* https://www.theguardian.com/australia-news/2017/feb/13/scott-morrison-and-ray-hadley-laugh-about-coal-prop-great-s.

Hunter, B. (2004). *The greenpeace to Amchitka.* Vancouver: Arsenal Pulp Press.

ICEMS. (2012, November 4). ICEMS position paper on the cerebral tumor court case. http://icems.eu/docs/ICEMS_Position_paper.pdf?f=/c/a/2009/12/15/MNHJ1B49KH.DTL.

International Energy Agency. (2009). *Gadgets and gigawatts: Policies for energy efficient electronics—Executive summary.* Paris: Organization for Economic Cooperation and Development.

IPCC. (2014). Climate change 2014 synthesis report—Summary for policy makers. *Intergovernmental panel on climate change.* Available at: https://www.ipcc.ch/pdf/assessment-report/ar5/syr/AR5_SYR_FINAL_SPM.pdf.

Isis1174. (2014, August 24). Leonardo DiCaprio ALS ice bucket challenge [Video file]. YouTube. Retrieved from https://www.youtube.com/watch?v=VwCgCrkdZAo.

Itzkoff, D. (2009, December 22). Opening Pandora's box: The Arguments Over 'Avatar'. *New York Times.* http://artsbeat.blogs.nytimes.com/2009/12/22/opening-pandoras-box-the-arguments-over-avatar/?_r=0.

Jackson, T. (2010). *Prosperity without growth? The transition to a sustainable economy.* London: Sustainable Development Commission.

Jackson, W. J., Argent, R. M., et al. (2017). *Australian state of the environment 2016: Overview.* Canberra: Australian Government Department of the Environment and Energy.

Jamieson, A. (2017, March 24). Keystone XL pipeline: Trump issues permit to begin construction. *The Guardian.* https://www.theguardian.com/environment/2017/mar/24/keystone-xl-pipeline-permit-trump-administration.

Jeffries, A. (2017, March 5). Google's featured snippets are worse than fake news. *The Outline.* https://theoutline.com/post/1192/google-s-featured-snippets-are-worse-than-fake-news.

Jhally, S. (2006). Advertising at the edge of the apocalypse. In S. Jhally (Ed.), *The spectacle of accumulation.* New York: Peter Lang.

Jones, P. D., & Mann, M. (2004). Climate over past millennia. *Review of Geophysics,42,* 1–42.

Kalhoefer, K. (2017). How broadcast networks covered climate change in 2016. *Media Matters for America.* https://mediamatters.org/research/2017/03/23/how-broadcast-networks-covered-climate-change-2016/215718.

Kasser, T. (2002). *The high price of materialism.* Cambridge MA: MIT Press.

Katz-Rosene, R. (2017). From narrative of promise to rhetoric of sustainability: A genealogy of oil sands. *Environmental Communication, 11*(3), 401–414. doi: 10.1080/17524032.2016.1253597.

Kellner, D. (2004). *Media culture and the triumph of the spectacle.* Razón y Palabra, 39. http://www.razonypalabra.org.mx/anteriores/n39/dkelner.html.

Kellner, D. (2015). *Media spectacle and the crisis of democracy: Terrorism, war, and election battles.* New York: Routledge.

Ker, P., & Morton, A. (2012). Carbon fails to slow coal boom. *The Age,* 5.

Kirsch, S. (2010). Sustainable mining. *Dialectical Anthropology, 34*(1), 87–93.

Klein, B. (2009). *As heard on TV: Popular music in advertising.* Surrey: Ashgate.

Klein, N. (2014). *This changes everything: Capitalism vs the climate.* Toronto: Knopf Canada.

Koring, P., & Cryderman, K. (2013, September 10). 'Fort McMurray is a wasteland': Neil Young slams oil patch, Keystone plans. *The Globe and Mail.* http://www.theglobeandmail.com/news/national/fort-mcmurray-is-a-wasteland-neil-young-slams-oil-patch-keystone-plans/article14214514.

Kraus. (2011, September 26). Activist communique: Ottawa tar sands action call-out. Rabbble.ca. http://rabble.ca/blogs/bloggers/krystalline-kraus/2011/08/activist-communique.

Kretchmer, S. B. (2004). Advertainment: The evolution of product placement as a mass media marketing strategy. *Journal of Promotion Management, 10*(1/2), 37–55.

Kroes, N. (2011). Data is the new gold. *Europa.eu.* Available at: http://europa.eu/rapid/press-release_SPEECH-11-872_en.htm.

Kuehr, R., & Williams, E. (Eds.). (2003). *Computers and the environment: Understanding and managing their impacts.* Dordrecht: Kluwer Academic Publishers.

Kunkel, F. (2014). Daring deal. *Government Executive.* http://www.govexec.com/magazine/features/2014/07/daring-deal/88207/. Accessed 9 July 2014.

Lamphier, G. (2013, September 12). Lamphier: No need to listen to Neil Young's take on the oilsands. *Edmonton Journal.* http://www.edmontonjournal.com/business/Lamphier+need+listen+Neil+Young+take+oilsands/8900746/story.html.

Landers, K. (2016, December 22). Canavan 'confident' Adani project to happen and 'deliver billions in taxes. *ABC Radio: AM.* http://www.abc.net.au/am/content/2016/s4596224.htm.

Landry, M. L. (2015). Speech by House of Representatives for Capricornia. Queensland at House of parliaments, Hansard, 20th August 2015.

Le Billion, P., & Carter, A. (2012). Securing Alberta's tar sands: Resistance and criminalization on a new energy frontier. In M. A. Schnurr & Larry A. Swatuk (Eds.), *Natural resources and social conflict: Towards critical environmental security* (pp. 170–192). New York: Palgrave Macmillan.

Levant, E. (2010). *Ethical oil: The case for canada's oil sands.* Toronto: McClelland and Stewart.

Levant, E. (2014, January 14). *LEVANT: Hypocrite Neil. Facebook post.* https://www.facebook.com/permalink.php?id=131404586952777&story_fbid=10152846353205260. Accessed 24 July 2017.

Lewis, J., Williams, A., & Franklin, B. (2008). Four rumours and an explanation: A political economic account of journalists. Changing newsgathering and reporting practices. *Journalism Practice, 2*(1), 27–45.

Lewis, J. (2013). *Beyond Consumer Capitalism: Media and the limits to imagination.* London: Polity.

Lewis, J. (2015). Cycle of silence: the strange case of disappearing environmental issues. *The Conversation.* https://theconversation.com/cycle-of-silence-the-strange-case-of-disappearing-environmental-issues-36306.

Lewis, J. (2016). Quick and dirty news: The prospect for more sustainable journalism. In P. Berglez, U. Olausson, & M. Ots (Eds.), *Sustainable journalism.* New York: Peter Lang.

Li, S. (2002). Junk-buyers as the linkage between waste sources and redemption depots in urban China: The case of Wuhan. *Resources, Conservation and Recycling, 36*(4), 319–335.

Linzner, R., & Salhofer, S. (2014). Municipal solid waste recycling and the significance of informal sector in urban China. *Waste Management and Research, 32*(9), 896–907.

Liu, Y., Li, X., Xu, Y. (2008). The planning of recycling system in Beijing. *Environment and Sustainable Development (in Chinese),* 1–2.

Ludlow, M. (2015). We're losing the PR battle, says BHP's Henry. *Australian Financial Review.* http://www.afr.com/business/mining/were-losing-the-pr-battle-says-bhp-coal-boss-20150917-gjpi48.

Lynch, J., & McGoldrick, A. (2005). *Peace journalism.* Stroud, UK: Hawthorn.

Lynham, A. (2016). Record of proceedings. Queensland Parliament. 15th March 2016.

Macdonald, I. (2016). Speech at Queensland Parliament, May 2, 2016.

Maher, S. (2011). Labor vows to fight for carbon plan. *The Australian,* 1.

Malm, A. (2016). *Fossil capital: The rise of steam power and the roots of global warming.* London: Verso.

Mann, M., Bradley Raymond, S. B., & Hughes, M. (1998). Global scale temperature patterns and climate forcing over the past six centuries. *Nature, 392,* 779–787.

Manning, P. (2009). Advertising aside, coal sector hasn't dug deep. *The Age,* 2.

Manyika, J., et al. (2015). *Unlocking the potential of the internet of things.* New York: McKinsey. http://www.mckinsey.com/business-functions/business-technology/our-insights/the-Internet-of-things-the-value-of-digitizing-the-physical-world.

Martinelli, D. K. (2011). Political public relations: Remembering its roots and classics. In J. Strömbäck & S. Kiousis (Eds.), *Political public relations: Principles and applications* (pp. 33–54). New York: Routledge.

Maude, F. (2012). Francis Maude speech to policy exchange—The big data opportunity. Available at. https://www.gov.uk/government/speeches/francis-maude-speech-to-policy-exchange-the-big-data-opportunity.

Maxwell, R., & Miller, T. (2012). *Greening the media.* Oxford: Oxford University Press.

Maxwell, R. (2015a). Social liabilities of digitising cultural institutions: Environment, labor, waste. In J. O'Connor & K. Oakley (Ed.), *The Routledge companion to the cultural industries* (pp. 392–440). London: Routledge.

Maxwell, R. (Ed.). (2015b). *The Routledge companion to labor and media*. New York: Routledge.

Maxwell, R., & Miller, T. (2014, July 2). Don't be fooled by the abundance of green apps. *Psychology Today* https://www.psychologytoday.com/blog/greening-the-media/201407/don-t-be-fooled-the-abundance-green-apps.

Maxwell, R., & Miller, T. (2016a). The propaganda machine behind the controversy over climate science: Can you spot the lie in this title? *American Behavioral Scientist, 60*(3), 288–304. Originally published online, October 29, 2015. http://journals.sagepub.com/doi/pdf/10.1177/0002764215613405.

Maxwell, R., & Miller, T. (2016b). Cancer and cellphones in the news—It's complicated. *Psychology Today.* https://www.psychologytoday.com/blog/greening-the-media/201606/cancer-and-cellphones-in-the-news-it-s-complicated. Accessed 19 Jan 2017.

Mayer-Schönberger, V., & Cukier, K. (2013). *Big Data: A revolution that will transform how we live, work and think*. London: John Murray.

McCarthy, J. (2014). Abbott sold on coal. *The Courier Mail*, 24.

McChesney, R., & Nichols, J. (2010). *The death and life of american journalism: The media revolution that will begin the world again*. Nation Books.

McCurdy, P. (2013). Conceptualising celebrity activists: The case of Tamsin Omond. *Celebrity Studies, 4*(3), 311–324.

McCurdy, P. (2011). Theorizing "Lay Theories of Media": A case study of the dissent! Network at the 2005 Gleneagles G8 summit. *International Journal of Communication, 5*, 619–638. Available from: http://ijoc.org/index.php/ijoc/article/view/842/559.

McCurdy, P. (2013). Conceptualising celebrity activists: The case of Tamsin Omond. *Celebrity Studies, 4*(3), 311–324.

McDermott, V. (2014). CAPP reacts to DiCaprio, Aronofsky visit. *Fort McMurray Today*. Retrieved August 25, 2014 from http://www.fortmcmurraytoday.com/2014/08/25/capp-reacts-to-dicaprio-aronofsky-visit.

McGoldrick, A., & Lynch, J. (2014). Audience responses to peace journalism. *Journalism Studies*. doi:10.1080/1461670X.2014.992621.

McGuire, P. (2014, January 24). 'Neil Young Lies' is pro-oil propaganda at its worst. *Vice Magazine*. http://www.vice.com/en_ca/read/neil-young-lies-is-pro-oil-propaganda-at-its-worst.

McKendrick, J. (2013). In the rush to cloud computing, here's one question not enough people are asking. *Forbes*. https://www.forbes.com/sites/joemckendrick/2013/02/19/in-the-rush-to-cloud-computing-heres-one-question-not-enough-people-are-asking/#536cb5cf7194. Accessed 26 June 2017.

McKendrick, J. (2014). We're all outsourcers now, thanks to cloud. *Forbes.* https://www.forbes.com/sites/joemckendrick/2014/08/11/were-all-out-sourcers-now-thanks-to-cloud/#1fb2448467aa. Accessed 26 June 2017.

McKenzie, N., Baker, R., & Ker, P. (2017, February 15). The coal war: inside the fight against Adani's plane to build Australia's biggest coal mine. *The Sydney Morning Herald.* http://www.smh.com.au/business/mining-and-resources/the-coal-war-inside-the-fight-against-adanis-plans-to-build-australias-biggest-coal-mine-20170213-gubn21.html.

McKibben, B. (2011, September 13). Tar-sands showdown: The fight over the future of energy. *Wired.* https://www.wired.com/2011/09/bill-mckibben-tar-sands-qa/.

McKibben, B. (2012) *Global warming's terrifying new math, Rolling Stone.* Access at http://www.rollingstone.com/politics/news/globalwarmings-terrifying-new-math-20120719.

Mckie, R. (2012, March 3). Death threats, intimidation and abuse: climate change scientist Michael E. Mann counts the costs of honesty. *The Guardian.* (https://www.theguardian.com/science/2012/mar/03/michael-mann-climate-change-deniers.

McKnight, D., & Hobbs, M. (2013). Public contest through the popular media: The mining industry's advertising war against the Australian Labor government. *Australian Journal of Political Science, 48*(3), 307–319.

McInerny, T. K. (2013). Letter to federal communications commissioners Mignon L. Clyburn and Margaret A. Hamburg. *American Academy of Pedriatrics.* https://ecfsapi.fcc.gov/file/7520941318.pdf. Accessed 19 Jan 2017.

Meade, A. (2015). Coal not so "amazing", public say, as mining industry advertising backfires. *The Guardian.* https://www.theguardian.com/environment/2015/nov/06/coal-not-so-amazing-public-say-as-mining-industry-advertising-backfires.

Meadows, D. H., et al. (1972). *The limits to growth: A report for the club of rome's project on the predicament of mankind.* New York: Universe Books.

Medina, M. (2010). *Solid waste, poverty and the environment in developing country cities—Challenges and opportunities.* United Nations University.

Meister, M. (2015). Celebrity culture and environment. In A. Hansen & R. Cox (Eds.), *The Routledge handbook of environment and communication* (pp. 281–289). London: Routledge.

Merritt, D. (1995). *Public Journalism and Public Life: Why telling the news is not enough.* Hillsdale, NJ: Lawrence Erlbaum Associates.

Met Office. (2017, January 18). 2016: One of the warmest two years on record. http://www.metoffice.gov.uk/news/releases/2017/2016-record-breaking-year-for-global-temperature.

Meyer, D. S., & Gamson, D. (1995). The challenge of cultural elites: Celebrities and social movements. *Sociological Inquiry, 65*(2), 181–206.

Michel-kerjan, E. (2013). Harnessing financial innovation to strengthen disaster resilience, PricewaterHouseCoopers. Available at: http://www.pwc.com/Fgovernance-risk-compliance-consulting-services/Fresilience/Fpublications/Fpdfs/Fissue3/Fharnessing_financial_innovation.pdf.

Miller, B. (2012). *Marketplace advocacy campaigns: Generating public support for business and industry.* New York: Cambria Press.

Miller, B., & Sinclair, J. (2009). Community stakeholder responses to advocacy advertising. *Journal of Advertising,38*(2), 37–51.

Mills, M. V. (2013). *The cloud begins with coal.* Washington, D.C.: National Mining Association.

Milman, O. (2017, March 9). EPA head Scott Pruitt denies that carbon dioxide causes global warming. *The Guardian.* https://www.theguardian.com/environment/2017/mar/09/epa-scott-pruitt-carbon-dioxide-global-warming-climate-change.

Monbiot, G. (2009). If you want to know who's to blame for Copenhagen, look to the U.S. Senate. http://www.theguardian.com/commentisfree/2009/dec/21/copenhagen-failure-us-senate-vested-interests.

Moore, J. W. (2015). *Capitalism in the web of life: Ecology and the accumulation of capital.* London: Verso.

Moore, J. W. (Ed.). (2016). *Anthropocene or capitalocene? Nature, history, and the crisis of capitalism.* Oakland, CA: PM Press.

Mosco, V. (2014). *To the cloud: Big data in a turbulent world.* Boulder: Paradigm.

Murdock, G. (2013). Communication in common. *The International Journal of Communication, 7,* 154–172.

Murdock, G. (2016). Reimaging the cultural commons. In K. Mitschka & K. Uterberger (Eds.), *Public social value* (pp. 129–146). Vienna: Osterreichischer Rudfunk, ORF.

Murphy, K. (2016). Liberal party donations to be scrutinised by Senate inquiry. *The Guardian.* https://www.theguardian.com/australia-news/live/2016/apr/19/australia-begins-its-unofficial-election-campaign-politics-live.

Nakashima, E. (2015). Hacks of OPM databases compromised 22.1 million people. Federal authorities say. *Washington Post.* https://www.washingtonpost.com/news/federal-eye/wp/2015/07/09/hack-of-security-clearance-system-affected-21-5-million-people-federal-authorities-say/. Accessed 9 July 2015.

National Cancer Institute. (nd.). Cell phones and cancer risk. https://www.cancer.gov/about-cancer/causes-prevention/risk/radiation/cell-phones-fact-sheet. Accessed 19 Jan 2017.

National Resources Defense Council. (2016). *Annual Report 2015.*

National Snow and Ice Data Centre. (2017, March 22). Arctic sea ice news and analysis: Arctic sea ice maximum at record low for third straight year. http://nsidc.org/arcticseaicenews/2017/03/arctic-sea-ice-maximum-at-record-low/.

Nease, K. (2008, October 9). Scream actress Neve Campbell visits oilsands developments: 'I'm horrified,' she says. *Fort McMurray Today*. Retrieved July 24, 2016 from http://www.fortmcmurraytoday.com/2008/10/09/scream-actress-neve-campbell-visits-oilsands-developments-im-horrified-she-says.

Neil Young Lies. (2016). Neil young lies. *Ethical Oil*. http://neilyounglies.ca/.

Nerman, D. (2015, December 10). Leonardo DiCaprio's chinook climate change comments mocked by Alberta politicos on Twitter. *Calgary Eyeopener*. http://www.cbc.ca/news/canada/calgary/leonardo-dicaprio-chinooks-climate-change-1.3358972.

Neyrat, F. (2015). Economy of Turbulence: How to Escape from the Global State of Emergency? *Philosophy Today, 59*(4), 657–669.

NRDC. (2017). NRDC annual report 2015. https://www.nrdc.org/sites/default/files/nrdc-annual-report-2015.pdf. Accessed 19 Jan 2017.

Oreskes, N., & Conway, E. M. (2010). *Merchants of doubt: How a handful of scientists obscured the truth on issues from tobacco smoke to global warming*. Bloomsbury Press.

Oreskes, N., & Conway, E. M. (2011). *Merchants of doubt: How a handful of scientists obscured the truth on issues from tobacco smoke to global warming*. Boston: Bloomsbury Publishing USA.

Orr, G. (2007). Political disclosure regulation in Australia: Lackadaisical law. *Election Law Journal, 6*(1), 72–88.

Oxford Dictionaries. (2016). Word of the year 2016. Available at https://en.oxforddictionaries.com/word-of-the-year/word-of-the-year-2016.

Packard, V. (1960). *The waste makers*. Harmondsworth: Penguin.

Palaszczuk, A. (2016). Speech by Queensland Premier Anastazia Palaszczuk in the Queensland Assembly on 19 April 2016.

Parenti, C. (2011). *Tropic of chaos: Climate change and the new geography of violence*. Nation books.

Parkhill, D. F. (1966). *The challenge of the computer utility*. Reading, MA: Addison-Wesley.

Pearse, G. (2009). Quarry vision: Coal, climate change and the end of the resources boom. *Quarterly Essay, 33*, 1.

Pearse, G., McKnight, D., & Burton, B. (2013). *Big coal: Australia's dirtiest habit*. Sydney: A NewSouth Book.

Pembina. (2012). *Clearing the air on oilsands emissions*. Retrieved June 20, 2016, from http://www.pembina.org/pub/2393.

Pfau, M., & Wan, H. (2006). Persuasion: An intrinsic function of public relations. In C. Botan & V. Hazleton (Eds.), *Public Relations Theory II* (pp. 101–136). London: Routledge.

Polden, D. (2011, October). Mass arrests at tar sands protest. *Peace News.* Issue 2538. http://peacenews.info/node/6329/mass-arrests-tar-sands-protest.

Potter, W. (2011). *Green is the new red: An insider's account of a social movement under siege.* City Lights Publishers.

Powles, J. (2015). Internet of things: The greatest mass surveillance infrastructure ever. *The Guardian.* https://www.theguardian.com/technology/2015/jul/15/Internet-of-things-mass-surveillance. Accessed 15 July 2015.

Priest, M. (2011). Miners dig deep to battle carbon tax. *Australian Financial Review,* 1.

Priest, M., & Walsh, K. (2012). ACCC: third-party's error no defence. *Australian Financial Review.*

Prime Minister Transcript. (2014). Retrieved from http://pmtranscripts.pmc.gov.au/release/transcript-23795.

Qiu, J. L. (2016a). Locating worker-generated content (WGC) in the world's factory. In R. Maxwell (Ed.), *The Routledge companion to labor and media* (pp. 303–314). New York: Routledge.

Qiu, J. L. (2016b). *Goodbye iSlave.* University of Illinois Press.

Rams, D. (2015). Agbogbloshie: Ghana's 'trash world' may be an eyesore—But it's no dump. *Ecologist.* http://www.theecologist.org/News/news_analysis/2959811/agbogbloshie_ghanas_trash_world_may_be_an_eyesore_but_its_no_dump.html. Accessed 20 Jan 2017.

Randalls, S. (2010). Weather profits: Weather derivatives and the commercialization of meteorology. *Social Studies of Science, 40*(5), 705–730.

Readfearn, G. (2014). What does Australian Prime Minister Tony Abbott really think about climate change? *The Guardian: Australian Edition.* https://www.theguardian.com/environment/planet-oz/2014/jun/16/what-does-australian-prime-minister-tony-abbott-really-think-about-climate-change.

Real-25-hour. (2012). Billionaire Waste Village. RealMedia (真实传媒).

Recycling Today. (2015). Chinese city moves electronics recycling activities to industrial park. *Recycling Today.* http://www.recyclingtoday.com/article/ban-guiyu-industrial-park-visit/. Accessed 20 Jan 2017.

Restart Project. (2015). *The footprint of those iPhones.* https://therestartproject.org/consumption/the-footprint-of-those-iphones/.

Reuters. (2012, October 19). Italy court ruling links mobile phone use to tumor. *Reuters.* http://www.reuters.com/article/2012/10/19/us-italy-phones-idUSBRE89I0V320121019. Accessed 19 Jan 2017.

Risk.net. (2010). A new direction for weather derivatives. Available at: http://www.risk.net/energy-risk/feature/1652654/a-direction-weather-derivatives.

Rittel, H., & Webber, M. M. (1973). Dilemmas in a general theory of planning. *Policy Sciences, 4,* 155–169.

Rockström, J., Steffen, W., Noone, K., Persson, Å., Chapin, F. S. III, Lambin, E., et al. (2009a). Planetary boundaries: Exploring the safe operating space for humanity. *Ecology and Society, 14*(2). http://www.ecologyandsociety.org/vol14/iss2/art32.

Rockström, J. et al. (2009a). A safe operating space for humanity. *Nature, 461* (7263), 472–475. Available at: http://dx.doi.org/10.1038/461472a.

Rockström, J., et al. (2009b). Planetary boundaries: Exploring the safe operating space for humanity. *Ecology and Society, 14*(2) (online, np.) http://www.ecologyandsociety.org/vol14/iss2/art32.

Rosen, J. (1991). Making journalism more public. *Communication, 12,* 267–284.

Rosenberg, J., & St. John III, B. (Eds). (2010). *Public Journalism 2.0: The promise and reality of a citizen-engaged press.* New York: Routledge.

Rowell, A. (2010, March 5). Welcome to Canada's AvaTAR sands. *Oil Change International.* http://priceofoil.org/2010/03/05/welcome-to-canadas-avatar-sands/.

Saad, L. (2017, March 14). Global warming concern at three-decade high in US. *Gallup.* http://www.gallup.com/poll/206030/global-warming-concern-three-decade-high.aspx.

Schauer, T. (2003). *The sustainable information society: Vision and risks.* Vienna: European Support Centre of the Club of Rome.

Scholz, Trebor, & Schneider, Nathan (Eds.). (2016). *Ours to hack and to own: The rise of platform cooperativism, a new vision for the future of work and a fairer internet.* New York: OR Books.

Schwab, K. (2016). The fourth industrial revolution: What it means, how to respond. *World Economic Forum.* Available at: https://www.weforum.org/agenda/2016/01/the-fourth-industrial-revolution-what-it-means-and-how-to-respond/.

SCOR Global. (2012). The transfer of weather risk faced with the challenges of the future. *Technical Newsletter June 2012.* Available at: http://www.scor.com/images/stories/pdf/library/newsletter/pc_nl_cata_naturelles_en.pdf.

Searchinger, T., & Heimlich, R. (2015). *Avoiding bioenergy competition for food crops and land.* Working Paper. Installment 9 of "Creating a Sustainable Food Future". Washington, DC: World Resources Institute.

Shang, Y. (2006). Research on status and development of Beijing's recycle system. *Journal of Beijing Institute of Finance and Commerce Management (in Chinese), 22,* 29–33.

Silverstone, R. (2007). *Media and morality: On the rise of the mediapolis.* Cambridge: Polity.

Slezak, M. (2016). Fossil-fuel industry gets $2,000 in 'subsidies' for each $1 in party donations. *The Guardian.* https://www.theguardian.com/environment/2016/feb/17/fossil-fuel-industry-gives-37m-to-major-parties-and-gets-big-subsidy-in-return.

Shelley, M. (1992 [1818]). *Frankenstein or the modern Prometheus.* London: Penguin Books.

Smith, J. (2011, September 21). Protesters? Check! Police? Check! Celebs? Also check! *Ottawa Metro.* http://www.metronews.ca/news/ottawa/2011/09/21/protesters-check-police-check-celebs-also-check.html.

Smith, A., & MacKinnon, J. (2007). *The 100-mile diet: A year of local eating.* Toronto: Random House Canada.

Speedwell Weather. (nd). *A quick guide to weather derivatives.* Available at: http://www.speedwellweather.com/PDF/Consultancy/A%20Quick%20Guide%20to%20Weather%20Derivatives.pdf.

Stand with Standing Rock. (2017). http://standwithstandingrock.net. Accessed 3 April 2017.

Steckel, J. C., Edenhofer, O., & Jakob, M. (2015). Drivers for the renaissance of coal. *Proceedings of the National Academy of Science, 112*(29), E3775–E3781.

Steffen, W., Crutzen, P. J., & McNeill, J. R. (2007, December). The Anthropocene: Are humans now overwhelming the great forces of nature? *Ambio, 36*(8), 614–621.

Steffen, W., Grinevald, J., Crutzen, P., & McNeil, J. (2011). The Anthropocene: conceptual and historical perspectives. *Philosophical Proceedings of the Royal Society, 369,* 842–867.

Steffen, W., Richardson, K., Rockström, J., Cornell, S. E., Fetzer, I., Bennett, E. M., et al. (nd). Planetary boundaries—an update. Stockholm Resilience Centre, Stockholm University. http://www.stockholmresilience.org/research/research-news/2015-01-15-planetary-boundaries—an-update.html.

Stein, L. D., et al. (2015). *Data Analysis: Create a Cloud Commons, 523*(7559). http://www.nature.com/news/data-analysis-create-a-cloud-commons-1.17916. Accessed 8 July 2015.

StEP. (2016). Guiding principles to develop e-waste management systems and legislation. *White Paper.* http://www.step-initiative.org/files/step-2014/Publications/Green%20and%20White%20Papers/Step_WP_WEEE%20systems%20and%20legislation_final.pdf.

Sterne, J. (2007). Out with the trash: On the future of new media. In C. R. Acland (Ed.), *Residual Media* (pp. 16–31). Minneapolis, MN: University of Minnesota Press.

Stiglitz, J. (2012). *The price of inequality.* London: Penguin.

Strasser, S. (2000). *Waste and want: A social history of trash.* New York: Holt.

Street, J. (2004). Celebrity politicians: Popular culture and political representation. *The British Journal of Politics and International Relations, 6*(4), 435–452.

Street, J. (2012). Do celebrity politics and celebrity politicians matter? *The British journal of politics and international relations, 14*(3), 346–356.

Stonebrook, S. (2014). 10 environmental nonprofit organizations that are changing the world. *Mother Earth News.* http://www.motherearthnews.com/nature-and-environment/environmental-policy/environmental-nonprofit-organizations-zmgz14amzsto. Accessed 19 Jan 2017.

Sullivan, B. (2015). The dirty cloud: IT will account for 12 percent of global electricity use by 2017. *TechWeek Europe.* http://www.techweekeurope.co.uk/e-innovation/greenpeace-data-centre-global-electricity-168134. Accessed 13 May 2015.

Sydney Morning Herald. (2014). Coal is 'good for humanity', says Tony Abbott at mine opening. Retrieved from http://www.smh.com.au/federal-politics/political-news/coal-is-good-for-humanity-says-tony-abbott-at-mine-opening-20141013-115bgs.html.

Sydney Morning Herald. (2016). Turnbull government eyes $1 billion Adani loan backed by new infrastructure. Available at http://www.smh.com.au/federal-politics/political-news/turnbull-government-eyes-1-billion-adani-loan-backed-by-new-infrastructure-fund-20161204-gt3joz.html.

Tackett, C. (2011, August 20). It begins. 70 arrested at white house on day 1 of tar sands action. *Tree Hugger.* http://www.treehugger.com/corporate-responsibility/it-begins-70-arrested-at-white-house-on-day-1-of-tar-sands-action-updated.html.

Tang, C., Feng, X. (2000). Stratification of rural migrant labor in Henan village. *Sociological Studies (in Chinese),* 72–85.

Tapley, I. (2015, December 4). Leonardo DiCaprio, Harvey Weinstein say 'Revenant,' 'Hateful Eight' productions witnessed climate change 'Firsthand'. *Variety.* http://variety.com/2015/film/in-contention/hateful-eight-revenant-climate-change-filming-1201653628/.

Tarbotton, R. (2010). James Cameron goes to the tar sands. *Huffington Post.* http://www.huffingtonpost.com/rebecca-tarbotton/james-cameron-goes-to-the_b_747551.html.

Tar Sands Action. (2011, August 20). 70 people arrested at white house to stop massive oil pipeline. *Tar Sands Action.* http://tarsandsaction.org/press/releases/aug20/.

Tarrow, S. (1998). *Power and movement: Social movements and contentious politics* [2nd ed.]. Cambridge: Cambridge University Press.

Taylor, C. & Meinshausen, M. (2014). Joint report to the land court of Queensland on "Climate Change—Emissions." *Brisbane.* Retrieved from http://envlaw.com.au/wp-content/uploads/carmichael14.pdf.

Turner, C. (2012). The Oil sands PR war: The down-and-dirty fight to brand Canada's oil patch. *Marketing Magazine.* Retrieved July 30, 2012 from http://www.marketingmag.ca/advertising/the-oil-sands-pr-war-58235.

Theoretical and Computational Biophysics Group, University of Illinois, Urbana-Champaign. (nd). Cryptochrome and magnetic sensing. http://www.ks.uiuc.edu/Research/cryptochrome/. Accessed 19 Jan 2017

The Great Barrier Reef Marine Park Authority. (2017, March 10). Second wave of mass bleaching unfolding on Great Barrier Reef. http://www.gbrmpa.gov. au/media-room/latest-news/coral-bleaching/2017/second-wave-of-mass-bleaching-unfolding-on-great-barrier-reef.

The Green Party of Canada. (2015). More than a round of applause for arts and culture. http://www.greenparty.ca/en/policy-background-2015/part-l. Accessed 19 Jan 2017.

The Stop the Adani Alliance. (2017). http://www.stopadani.com/about.

Thind, S. (2014). As temperatures tumble in North America, weather derivatives warm up. *Institutional Investor.* January 23, 2014. Available at: http://www. institutionalinvestor.com/Article/3300613/Asset-Management-Hedge-Funds-and-Alternatives/As-Temperatures-Tumble-in-North-America-Weather-Derivatives-Warm-Up.html#/.VjzJLyvYHR8.

Ting, I., & Begley, P. (2015). Political donations: Mining hits back at Labor. *Sydney Morning Herald.* http://www.smh.com.au/federal-politics/political-news/political-donations-mining-hits-back-at-labor-20150205-1372gf.html.

Touropia. (2016, November 7). 10 largest malls in the world. http:// www.touropia.com/largest-malls-in-the-world/.

Toxic Leaks. (2016, November 4). In Guiyu, the e-waste nightmare is far from over. *Toxic Leaks.* Available at https://toxicleaks.com/wiki/In_Guiyu,_the_e-waste_nightmare_is_far_from_over. Accessed 20 Jan 20 2017.

Tse, E., & Hendrichs, M. (2016). Well connected: The growing reach of China's internet sector. *South China Morning Post.* http://www.scmp.com/comment/ insight-opinion/article/1897072/well-connected-growing-reach-chinas-Internet-sector. Accessed 3 Jan 2016.

Tuchman, G. (1978). *Making news: A study in the construction of reality.* New York: Free Press.

Turner, C. (2012, July 30). The oil sands PR war: The down-and-dirty fight to brand Canada's oil patch. *Marketing Magazine.* http://www.marketingmag. ca/advertising/the-oil-sands-pr-war-58235.

UNESCO (2012). *Culture: A Driver and an Enabler of Sustainable Development.* Paris. http://www.un.org/millenniumgoals/pdf/Think%20Pieces/2_culture. pdf.

United Nations. (2014). Resolution adopted by the general assembly on 20 December 2013 (A/RES/68/223). *Culture and Sustainable Development.*

United States Environmental Protection Agency. (2017). 'Causes of Climate Change'. http://www.epa.gov/climatechange-science/causes-climate-change. Accessed 10 March 2017.

Vertatique. (2014). *The growth in global telecom GHG emissions.* http://www. vertatique.com/telecom-emissions-300-mtco2-and-growing.

Waltzer, H. (1988). Corporate advocacy advertising and political influence. *Public Relations Review, 14*(1), 41–55.

Wang, J., Han, L., et al. (2008). The collection system for residential recyclables in communities in Haidian District, Beijing: A possible approach for China recycling. *Waste Management, 28*(9), 1672–1680.

Wasko, J., Phillips, M., & Purdie, C. (1993). Hollywood meets madison avenue: The commercialization of US films. *Media, Culture and Society, 15*(2), 271–293.

Watson, I. (2013). China: The electronic wastebasket of the world. *CNN.* http://www.cnn.com/2013/05/30/world/asia/china-electronic-waste-e-waste.

Wearden, G. (2016, January 20). Leonardo DiCaprio savages corporate greed of big oil: 'Enough is enough'. *The Guardian.* https://www.theguardian.com/business/2016/jan/20/leonardi-dicaprio-savages-corporate-greed–big-oil-enough-is-enough.

Webster, F. (2006). *Theories of the information society* (3rd ed.). London: Routledge.

Weinberg, A. S., Pellow, D. N., & Schnaiberg, A. (2000). *Urban recycling and the search for sustainable community development.* Princeton: Princeton University Press.

Weiss, P. (2002). *Borders in cyberspace: Conflicting public sector information policies and their economic impacts.* U.S. National Weather Service. Available at: http://www.nws.noaa.gov/sp/Borders_report.pdf.

Wheeler, M. (2013). *Celebrity politics.* Polity.

Whiteman, G., Hope, C., & Wadhams, P. (2013, July 25). Vast costs of arctic change. *Nature, 499,* 401–403.

Wiener, N. (1948). *Cybernetics; or, control and communication in the animal and the beast.* New York: Wiley.

Wihbey, J. (2011, October 11). A mixed media verdict on tar sands mass action, arrests. *Yale Climate Connections.* http://www.yaleclimateconnections.org/2011/10/tar-sands-mass-action-arrests/.

Wilkinson, R., & Pickett, K. (2009). *The spirit level: Why equality is better for everyone.* London: Penguin.

Williams, E. (2011). Environmental effects of information and communications technologies. *Nature, 479,* 354–358.

Wilson, H. J. (2013). Wearables in the workplace. *Harvard Business Review.* September. https://hbr.org/2013/09/wearables-in-the-workplace.

Wilt, J. (2015, September 21). Celebrities and the oilsands: Help or hindrance?" *Desmog.* http://desmog.ca/2015/09/21/celebrities-and-oilsands-help-or-hindrance.

Wingrove. (2010a, September 30). Q&A James Cameron talks oil sands with the globe. *The Globe and Mail.* Retrieved July 24, 2016, from http://www.theglobeandmail.com/news/national/james-cameron-talks-oil-sands-with-the-globe/article1380446/.

Wingrove, (2010b, October 22). Syncrude to pay $3M for duck deaths. *The Globe and Mail.* Retrieved July 24, 2016, from http://www.theglobeandmail.com/report-on-business/industry-news/energy-and-resources/syncrude-to-pay-3m-for-duck-deaths/article4085700/.

WorldBank. (1999). *Green industry. New Roles for Communities, Markets, and Government.* Oxford: Oxford University Press.

World Economic Forum. (2016). *The new plastics economy: Rethinking the future of plastics.* Geneva: World Economic Forum. https://www.weforum.org/reports/the-new-plastics-economy-rethinking-the-future-of-plastics.

World Meteorological Organisation. (2017). *WMO Statement on the State of the Global Climate in 2016.* Geneva: World Meteorological Organisation.

Wratten, A. (2009). Climate change is hot topic. *The Morning Bulletin.*

WRMA, n.d. (a). History of the weather market. *Weather Risk Management Association.* Available at: http://web.archive.org/web/20131210021948/, http://wrma.org/risk_history.html.

WRMA, n.d. (b). Trading weather risk. *Weather Risk Management Association.* Available at: http://www.wrma.org/risk_trading.html.

WRMA. (2009). Weather market in Asia, Europe grew in 2008–2009 while overall market declined. *Weather Risk Management Association.* Available at: http://www.wrma.org/pdf/WRMAPwC2009Surveypressrelease.pdf.

WRMA. (2011). Weather derivatives market shows robust growth in 2010–2011. *Weather Risk Management Association.* Available at: http://web.archive.org/web/20131203203136/, http://www.wrma.org/pdf/WRMA2011Industry SurveypressreleaseFINAL.pdf.

Xiang, W.-N. (2013). Working with wicked problems in socio-ecological systems: Awareness, acceptance, and adaptation. *Landscape and Urban Planning, 110,* 1–4.

Xiang, W.-N., Stuber, R. M. B., & Meng, X. (2011). Meeting critical challenges and striving for urban sustainability in China. *Landscape and Urban Planning, 100,* 418–420.

Yang, Z., Zhang, Y., Li, S., Liu, H., Zheng, H., Zhang, J., et al. (2014). Characterizing urban metabolic systems with an ecological hierarchy method, Beijing, China. *Landscape and Urban Planning, 121,* 19–33.

Yuan, Z., Bi, J., & Moriguichi, Y. (2006). The circular economy: A new development strategy in China. *Journal of Industrial Ecology, 10,* 4–8.

Zalasiewicz, J., Williams, M., Haywood, A., & Ellis, M. (2011). The Anthropocene: A new epoch of geological time? *Philosophical Transactions of the Royal Society, 369,* 835–841.

Zelko, X. (2013). *Make it a green peace!: The rise of a countercultural environmentalism.* Oxford: Oxford University Press.

Zhang, Q., Jiang, X., et al. (2017). Transboundary health impacts of transported global air pollution and international trade. *Nature, 543,* 705–709.

Zotos, G., Karagiannidis, A., Zampetoglou, S., Malamakis, A., Antonopoulos, I. S., Kontogianni, S., et al. (2009). Developing a holistic strategy for integrated waste management within municipal planning: Challenges, policies, solutions and perspectives for Hellenic municipalities in the zero-waste, low-cost direction. *Waste Management, 29,* 1686–1692.

Zou, Y., Wang, Y., Zhang, Y., & Ja-Koo, K. (2017). Arctic sea ice, Eurasia snow, and extreme winter haze in China. *Science Advances, 3.* http:// advances.sciencemag.org/content/3/3/e1602751.

Zwartz, H. (2016). Death in paradise: Family of Neil Lawrence still searching for answers nine months later. *The Sydney Morning Hearld.* http://www.smh.com. au/national/death-in-paradise-family-of-neil-lawrence-still-searching-for-answers-nine-months-later-20160301-gn79ke.html.

# INDEX

© The Editor(s) (if applicable) and The Author(s) 2017          249
B. Brevini and G. Murdock (eds.), Carbon Capitalism and Communication,
Palgrave Studies in Media and Environmental Communication,
DOI 10.1007/978-3-319-57876-7

Printed by Books on Demand, Germany